SAVING EARTH

Climate Change and the Fight for Our Future

OLUGBEMISOLA RHUDAY-PERKOVICH

Pictures by **Tim Foley** • Introduction by **Nathaniel Rich**

SAVING

EARTH

Climate Change and the Fight for Our Future

Farrar Straus Giroux • New York

Farrar Straus Giroux Books for Young Readers
An imprint of Macmillan Publishing Group, LLC
120 Broadway, New York, NY 10271 • mackids.com

Adapted, in part, from *Losing Earth*: *A Recent History* by Nathaniel Rich,
published by Farrar Straus Giroux, 2018

Our books may be purchased in bulk for promotional, educational, or business use.
Please contact your local bookseller or the Macmillan Corporate and Premium
Sales Department at (800) 221-7945 ext. 5442 or by email at
MacmillanSpecialMarkets@macmillan.com.

Library of Congress Cataloging-in-Publication Data

Names: Rhuday-Perkovich, Olugbemisola, author. | Foley, Tim, 1962- illustrator.
Title: Saving earth : climate change and the fight for our future / Olugbemisola
   Rhuday-Perkovich ; pictures by Tim Foley ; introduction by Nathaniel Rich.
Description: First edition. | New York : Farrar Straus Giroux, 2022. | Includes
   bibliographical references and index. | Audience: Grades 4-6 | Summary: "A timely and
   important illustrated nonfiction guide for middle grade readers about the history of our
   fight against climate change, and how young people today can rise to action"—Provided
   by publisher.
Identifiers: LCCN 2021025164 | ISBN 978-0-374-31305-0 (hardcover)
Subjects: LCSH: Climatic changes—Juvenile literature. | Green movement—
   Juvenile literature.
Classification: LCC QC903.15 .R45 2022 | DDC 363.738/74—dc23
LC record available at https://lccn.loc.gov/2021025164

Grunge texture © Miloje/Shutterstock; doodle page divider © mhatzapa/Shutterstock;
leaves border © hana honoka/Shutterstock; leaf © Gaianami Design/Shutterstock;
adhesive tape, paper, pin, and pushpin © TMvectorart/Shutterstock.

First edition, 2022
Book design by Angela Jun
Printed in the United States of America by LSC Communications, Harrisonburg, Virginia

ISBN 978-0-374-31305-0 (hardcover)
1 3 5 7 9 10 8 6 4 2

To my parents, Pat and Olu Shonubi; my grandparents, Verna and Charles Robertson; and my daughter, Adedayo: You remind me that the wisdom of "back home" and the knowledge of our ancestors is at the foundation of my hope for the future.

CONTENTS

Part III: Handling the Truth

Introduction

Nathaniel Rich

What follows is the story of our time. One of the most incredible things about it is that very few people know how it began (though *you* will, soon), and nobody knows how it will end (though *you*, decades from now, might). The most surprising thing about it, however, is that you and I and everybody we know are its main characters. As you read this story, remember that you are also its hero.

The story of climate change, and our response to it, is the story of modern civilization. Consider all the godlike powers we take for granted: refrigeration; indoor cooking; temperature control; modern medicine; motorcycles, cars, buses, helicopters, airplanes, spaceships; the internet; supermarkets and everything you find in them; laundry machines, dishwashers, lightbulbs, garage doors, sneakers, books, plastic, TikTok. Nearly any technology that shapes our lives relies upon the consumption of massive quantities of energy. For nearly all of human history, most of that energy has come from burning fossils. Eons' worth of dead things—dinosaurs, trilobites, plants, trees, moss, algae—decomposed, sunk into the Earth, and were dug up and burned by enterprising human beings. Like something out of a zombie movie, however, the Earth's dead things are now seeking their revenge, overheating the planet and sowing chaos. Our greatest source of power has yielded the most vicious threat our civilization has faced.

The story of climate change is also the story of human failure: of warnings ignored, of missed opportunities, and of malevolent actors, chief among them those who have profited from the extraction and manufacture of fossil fuels. As I write this introduction, in 2022, we are only beginning to rid ourselves of the disinformation campaign that was codified at the highest levels of the oil and gas industry in the late 1980s, just as major global climate policy was being proposed and drafted by the United States and the world's other largest nations. The disinformation campaign persists, however, as does the political division it caused.

We have been brought to this point by the nearsighted decisions of powerful people in positions of great responsibility. Anyone born in the twenty-first century will, at some point in their lives, come to realize that they have been dealt a bad hand. Those who are most responsible for getting us into this mess are the ones who will be least harmed by it. The most innocent—the youngest—will suffer the most. The perverse cruelty of climate change cuts along every social divide. It aggravates nearly every form of inequality in society. Those who face the sharpest discrimination—people of color, Indigenous communities, the poor, the infirm, the undereducated—are most severely threatened. Those who want most badly to defeat the problem often have the least political power to do so.

This grim history does not doom us to a grim future. A vast range of outcomes remains available, with an enormous gulf between best- and worst-case scenarios. And more than forty years after Rafe Pomerance, James Hansen, and Al Gore first tried to force the United States government to act, a new generation has seized leadership of the fight, with sudden, and dramatic, consequences.

In *Saving Earth*, Olugbemisola Rhuday-Perkovich explains how we got here, gives a vivid portrait of the critical moment in which we find ourselves, and introduces the young visionaries who have reshaped the public conversation in just a few years. The future, we know, will look nothing like the past. It will be warmer, yes, and the strains of climate change will harm us in ways both darkly predictable and unforeseen. But the struggle against the ravages of warming will also be led by a new cohort of activists, engineers, and thinkers. This new generation doesn't look like the previous ones, and it won't act like them either. Young people faced with the threat of climate change in the 2020s will not tolerate delays or make excuses. They won't be distracted by the old political tricks. They will understand that survival of the world as we know it will depend on a profound commitment to honesty, responsibility, and, above all, justice.

Saving Earth is their story—and yours.

Part I

WHERE WE ARE

Today, almost nine out of ten Americans do not know
that scientists agree, well beyond the threshold of consensus,
that human beings have altered the global climate through the
indiscriminate burning of fossil fuels.

—**Nathaniel Rich,** *Losing Earth: A Recent History*

Chapter 1
LETTING OLD HABITS DIE, HARD

Vision, Voices, and Vegetables

In 2009, New York City students at the Brooklyn New School started a campaign that later became part of an effort called Styrofoam Out of Schools. Their mission? Well, you can probably guess it from the name. But to be clear, they wanted to get plastic-foam lunch trays out of schools.

The students weren't worried that the trays were too flimsy or ugly or made out of a weird substance that wasn't found in nature. They weren't trying to be "different," even though the idea of using anything other than plastic-foam trays in cafeterias was very different. They were concerned that the trays, which were used for only a few minutes before they were tossed in the trash, were not biodegradable and stayed around for hundreds, if not thousands, of years.

Styrofoam is the trade name for polystyrene foam, a plastic material with many tiny air pockets trapped in its structure. Like many plastics, polystyrene is derived from petroleum. Plastic foam is used in food containers, packaging, insulation, and more because it is lightweight, inexpensive to produce, and excellent at keeping hot things hot and cold things cold. But it's also . . . well, plastic, and not biodegradable. That's just one of its problems. Not only is fossil fuel an ingredient, but the manufacturing process releases greenhouse gases that cause global warming (which can be said of the manufacturing process of just about any product), and it involves toxic chemicals linked to cancer. When the foam is heated, toxic chemicals can transfer from the plastic into the food and drink the containers were designed to protect. And finally, after the plastic foam has been used just once, it often ends up floating in waterways and oceans, where it breaks into tiny pieces that animals mistake for food, which can lead to their early death.

PETROLEUM: Another word for oil. A liquid that is naturally present in certain layers of rock beneath the Earth's surface. It can be removed and refined to produce fuels including gasoline, kerosene, and diesel oil. Like coal, petroleum is a *fossil fuel*, meaning that it is formed over millions of years from the remains of plants and animals deep within the Earth's crust.

At the time, over four million foam trays in New York City alone were being trashed every week, to be taken to landfills, where plastic foam can last for up to five hundred years, or burned in fiery incinerators, where it can produce harmful fumes. The "disposable" lunch trays didn't just disappear. They hung around, degrading into smaller and smaller pieces and potentially leaching toxic chemicals, polluting and poisoning our soil, water, marine life—our very selves. Gross, right? And this was a normal part of everyday life.

Yet when the students thought about it, they didn't want gross to be the norm. So they got together, they spoke out, they contacted government officials—they became activists— and within a year, all New York City public schools (more than sixteen hundred at the time) had set up "Trayless Tuesdays," replacing plastic-foam trays with paper containers on, yeah, Tuesdays. In 2014, New York City public schools eliminated plastic-foam trays entirely from their cafeterias and replaced them with compostable, sustainably produced, Earth-friendly ones. Today, Styrofoam Out of Schools is called Cafeteria Culture and works with youth to "achieve zero-waste schools, plastic-free waters, and climate-smart communities."

At schools like the Brooklyn New School, young activists didn't just focus on trays. They grew food on the school playground. They instituted a salad bar and a schoolwide recycling and composting program, and made sustainability a part of the school curriculum. They changed their behavior, they changed their culture, they transformed their way of life.

We Can All Be Superheroes

In the beginning, the mission of those young activists might have seemed unnecessary, wacky, even impossible. A lot of people probably told the students they were wasting their time. But those kids knew that it wasn't just about school lunch. They had done the research and learned that our garbage and how we dispose of it says a lot about who we are. And it was pretty clear to them that they did not want to be polluting, poisoning, life-stealing citizens whose behavior was threatening the future of the Earth more than any superhero-movie villain ever could. You probably don't want to be that either. Even though cartoon villains do get to do some cool-seeming stunts for a while, things never turn out well for those bad guys in the end. And they leave a lot of destruction and pain in their path along the way.

But it turns out that many of us have kind of been those bad guys for a while. And if we continue to live in ways that damage our air, our waters, our Earth—that threaten our everything—our story won't end well either.

It's time to write a new one.

YOUNG CLIMATE ACTIVISTS IN THEIR OWN WORDS

They keep talking about climate change being a matter of the future, but they forget that for people of the Global South, it is a matter of now.

—Vanessa Nakate,
Uganda, founder of Youth for Future Africa and Rise Up Movement

It is when Indigenous peoples come together that powerful things happen. Through building relationships and sharing ideas, we can start to gather under the rafters of our own whare* to bring to light our own dreams, rather than just coming together when our governments or the UN wants us to.

—**India Logan-Riley,** Ngāti Kahungunu ki Ngāti Hawea ki Whatuiapiti, Aotearoa (New Zealand), a Māori activist for Indigenous rights who participated in United Nations climate talks

*A Māori building

My Government has failed to take steps to regulate and reduce greenhouse gas emissions, which are causing extreme climate conditions. This will impact both me and future generations. My country has huge potential to reduce the use of fossil fuels, and because of the Government's inaction I approached the National Green Tribunal.

—**Ridhima Pandey,** India, who, at age nine, brought a court case against the Indian government for failing to keep its Paris Agreement promises

For young indigenous women, access to information and participation in public policy remain a challenge . . . It is time that the world hears our voice and the country recognizes indigenous women as equal rights holders.

—**Rayanne Cristine Maximo Franca,** Brazil, Indigenous Youth Network

I started my activism quite young—at 11. That was when I first heard about this thing called climate change. As a young girl in Samoa, a small island in the south Pacific, hearing the implications it had for my island scared me and jumpstarted my passion to do something about it . . . And feeling that you have a team, that you're not alone, that we're all in this together. It's not just one person yelling from outside the UN building or our parliament. And where there are mass numbers, there's power. Our slogan is: "We're not drowning. We're fighting."

—**Brianna Fruean,** Samoa, founding member, 350.Samoa, Future Rush; first youth ambassador of the Secretariat of the Pacific Regional Environment Programme

One lesson much too well learnt must be that something has to be done, not through the complicated negotiations and conference rooms of the [United Nations] but through our own small actions. We can find solutions right here. In the long run this may well be the vehicle through which solid decisions are taken on climate change issues.

—**Winnie Asiti,** Kenya, cofounder of African Youth Initiative on Climate Change

You've probably heard the somber predictions, the dire promises, the downright terrifying tales about global warming, the greenhouse effect, and climate change.

The basic science of the "greenhouse effect" is not especially complicated. It can be reduced to a simple statement: The more carbon dioxide in the atmosphere, the warmer the planet. And every year, by burning fossil fuels such as coal, oil, and gas, human beings belch increasingly obscene quantities of carbon dioxide into the atmosphere.

Carbon dioxide is an important part of the air we breathe on Earth. Made of one carbon and two oxygen atoms, CO_2 traps the energy and heat from the sun in our atmosphere. Without this and other greenhouse gases such as water vapor, heat would escape into space, and it would be too cold for us to survive. But too much carbon dioxide traps too much heat and warms the Earth to dangerous levels, contributing to climate change.

"Fixing" these truly worldwide problems might seem like impossible, superhero-level stuff. We're only human, right? But that's the thing. We are human. We are capable of changing our behavior and transforming the way we live. We've done it before. Maybe we can do it again. Maybe we can make the right changes to reduce our carbon footprint and address this global warming problem that threatens us all.

There are kids, regular kids like the Trayless Tuesday activists in New York City, kids like you, who think we can.

CAN YOU SEE YOUR CARBON FOOTPRINT?

Sort of. A *carbon footprint* is the total amount of greenhouse gases generated by a person's day-to-day, hour-to-hour actions. You can "see" it by recognizing how much you participate in the burning of fossil fuels in your daily life. Fossil fuels may sound like an ancient dinosaur energy drink that enabled T. rex to use those tiny hands to its advantage, but they are natural energy sources, like coal, petroleum, and natural gases, that formed from the remains of plants and animals over millions of years. Unlike solar and wind power, fossil fuels are *nonrenewable resources*; they cannot be replenished once they are used. The fossil fuels used to power the engines that move our vehicles, create our electricity, make our products, heat our homes, cook our food, and do many other everyday things, all contribute to our carbon footprint. So do the emissions created by growing and processing the food we eat. The scientists at NASA explain it in these simple terms: "Your carbon footprint is the amount of carbon dioxide released into the air because of your own energy needs." For example, walking or using public transportation instead of cars and using a fan instead of an air conditioner can reduce your carbon footprint because you'll burn less fossil fuel. Carbon footprint calculators are easy to find on the internet.

Another young person who believes in our ability to make the right changes is Isra Hirsi in the United States, who at sixteen organized U.S. participation in the International Youth Climate Strike. "These strikes are happening all over the world," says Hirsi. "Getting young people out, going to state capitols, going to city halls, going to the nation's capital and talking about these things, that says something. That's what we're trying to do: Change the conversation . . . Obviously, one strike isn't going to change everything, but this isn't the last strike."

It's not the end, not yet. Maybe we can make it a new beginning.

Chapter 2

WHAT IS CLIMATE CHANGE, AND WHERE DID IT COME FROM?

Climate Change vs. Global Warming: Who! Will! Win?

Settle down, it's not a battle. We often hear the terms used interchangeably, but they're two different things.

Climate change refers to the long-term change in temperature and weather patterns that happen in any one place, or across the world. Climate change does include the warming temperatures on Earth, but it also includes extreme weather events like hurricanes, and the loss of ice from melting glaciers that cause rising sea levels.

When we talk about climate change now, we're usually referring to the effects of *global warming*—the rising global temperatures that occur as a result of our increased burning of fossil fuels such as coal, oil, and gas, which releases carbon dioxide into the atmosphere.

The gargantuan threat of climate change should force us to re-think global systems that are disastrous for the planet and deeply inequitable. These systems mean 85 people in the world have more wealth and consume more than 3.5 billion people—half the world's population. Our survival is dependent on governments making binding and drastic commitments to reduce emissions. But it is also dependent on a commitment to finally deliver on human rights promises and provide Development Justice to all.

—**Alina Saba,** Nepal, cofounder, Nepal Policy Center

Carbon dioxide, water vapor, and methane are the main *greenhouse gases* in the Earth's atmosphere. Like the glass in a greenhouse's roof and walls, these gases let some radiant energy from the sun pass through to the Earth's surface; they absorb other radiant energy from both the sun and from the Earth's surface and in turn send radiant energy upward toward space as well as downward to Earth. This energy warms Earth's surface. In a loop effect, the warmer the Earth's surface gets, the more energy it releases back to the atmosphere and the more radiant energy the atmosphere then absorbs and sends back down toward Earth.

Although climate science skeptics may suggest that the scientific examination of global warming and its impact on climate is still in its infancy, it has actually been going on for a long, long time. Nearly seven decades ago, in 1956, one of the most widely read newspapers in the world, the *New York Times*, published this simple summary of the greenhouse effect in an article about research scientist Gilbert Plass:

> The atmosphere acts like the glass of a greenhouse. Solar radiation passes through to the earth readily enough, but the heat radiated by the earth is at least

partly held back. That is why the earth's surface is relatively warm. Carbon dioxide, water vapor and ozone all check radiation of heat. Of the three gases that check radiation, carbon dioxide is especially important . . . As the amount of carbon dioxide increases, the earth's heat is more effectively trapped, so that the temperature rises . . . Despite nature's way of maintaining the balance of gases the amount of carbon dioxide in the atmosphere is being artificially increased as we burn coal, oil, and wood . . . It will have a profound effect on our climate.

THERE'S NOTHING NEW UNDER THE SUN

Eunice Foote (1819–1888) was an amateur scientist and women's rights supporter. In the 1850s, Foote used an air pump, thermometers, and glass cylinders to discover that the heating effects of the sun were stronger in moist air than in dry air, and even stronger and more stubborn in the presence of carbon dioxide. CO_2, she found, trapped more heat than hydrogen, oxygen, or common air, and the cylinder containing carbon dioxide took much longer to cool down. Foote was years ahead of her time in theorizing that more carbon dioxide in the atmosphere would warm the planet. A male scientist presented Foote's research at the 1856 Annual Meeting of the American Association for the Advancement of Science, to little fanfare. Three years later, the now more well-known John Tyndall found similar results in his study of carbon dioxide.

John Tyndall (1820–1893) was an Irish math and science professor who researched how different gases in our atmosphere absorb the Earth's radiant heat, concluding that water vapor and carbon dioxide were the strongest absorbers. Publishing his findings in 1859, he described what today we call the greenhouse effect: "The atmosphere admits of the entrance of the solar heat, but checks its exit; and the result is a tendency to accumulate heat at the surface of the planet."

Svante Arrhenius (1859–1927), a Swedish physicist, was the first to construct a mathematical model to calculate how the amount of carbon dioxide in the atmosphere affects our climate. In 1896, he predicted that the doubling of carbon dioxide due to fossil fuel burning would increase global temperatures by three to four degrees Celsius.

Roger Revelle (1909–1991), an oceanographer, and Austrian geochemist *Hans Suess* (1909–1993) demonstrated in 1957 that carbon dioxide from fossil fuels was accumulating in the atmosphere. At the time many scientists believed oceans absorbed most of the CO_2 from burning fossil fuels, but Revelle and Suess showed that the chemical mix of seawater prevents it from retaining all the extra carbon dioxide. "The increase of atmospheric CO_2 from this cause is at present small," they wrote, "but may become significant during future decades if industrial fuel combustion continues to rise exponentially." In 1956, *Time* magazine summarized his groundbreaking work for its readers: "Since the start of the industrial revolution, mankind has been burning fossil fuel (coal, oil, etc.) and adding its carbon to the atmosphere as carbon dioxide. In 50 years or so this process . . . may have a violent effect on the earth's climate."

Charles David Keeling (1928–2005) was hired by Revelle to measure carbon dioxide in the atmosphere. In the pristine air of Antarctica and high atop Mauna Loa in Hawaii, Keeling established a stable baseline of atmospheric carbon dioxide. Just two years later, in 1960, he reported the baseline had risen. The graph showing the rise in carbon dioxide levels over the years is called the Keeling Curve. The measurements continue today.

Change Is Only Natural—Unless It's Not

Way back in 1824, French scientist Joseph Fourier pointed out that the Earth's atmosphere acted kind of like a glass box, similar to a greenhouse that traps heat from sunlight and protects a gardener's plants from colder temperatures outside. Our atmosphere keeps us from burning up under the heat of the sun during the day, and from freezing completely at night. In a Goldilocks kind of way, our atmosphere has trapped just enough heat to keep things not too hot and not too cold and more or less "just right."

We know that some climate change is natural. Changes in the Earth's orbit have moved the planet closer to the sun and farther away again, causing warming and cooling cycles. Volcanic eruptions have filled the atmosphere with particles that blocked sunlight and cooled the planet; others released gases that made things hotter.

But things have been going wacky for a while. The world has warmed more than one degree Celsius (about two degrees Fahrenheit) since 1880, during the Industrial Revolution. That seems like a long time ago, right? And one degree sounds pretty insignificant. We don't usually feel the difference in temperature when it changes by a degree or two.

But we're talking about changes in climate, not weather. Weather refers to short-term, day-to-day variations in the atmospheric conditions in a specific area, while climate describes what the weather is like over a long period in a specific area. And over the whole history of time as we know it, these changes have been happening pretty fast. But now things are happening faster than ever.

The Industrial Revolution literally made for big business with large-scale manufacturing and production. It improved the quality of life for some people, but it also dramatically increased the brutal enslavement of Africans.

As the textile industry grew in England, the appetite for cheap cotton was fed by the inhumane system of using Black people's forced labor to produce and pick the cotton in the British colonies. By the 1790s there were approximately 700,000 enslaved people in the United States. The invention of the cotton gin in 1807 meant that harvested cotton could be processed quickly by machine, encouraging plantation owners to produce more cotton. Along with other developments of the Industrial Revolution, like weaving machines and steamboat transport, the cotton gin drastically increased production—and also increased the output of carbon dioxide. Cotton became a major U.S. export, and the soaring profits aggressively increased demands on the land—and on the enslaved human beings forced to work that land. Fueled by the extreme violence and ruthless dehumanization of slavery, the increased production put more greenhouse gases in the atmosphere, creating more changes in our climate.

Although this period of industrialization has often been seen as the start of our current climate change trend, researchers are

THE INDUSTRIAL REVOLUTION: The period from about the mid-1700s to the early 1900s, when countries in the Western world, beginning with England, left their agricultural roots behind and became more industrialized. People found a new source of energy in carbon-packed fossil fuels such as coal, oil, and natural gas and invented powerful machines to take care of what used to be done by hand or animals.

exploring the idea that negative human impact on the Earth's climate began even before the Industrial Revolution.

For example, some now believe that the European assault on the land and lives of Indigenous communities in North and South America actually changed the Earth's climate. "The invasion of the Americas changed everything on the planet," writes Indigenous journalist Mark Trahant. "There was a death rate of more than 90 percent of the people . . . At that time, the Indigenous population of the Americas was 10 percent of the world's population and home to the largest and most complex cities."

Researchers at University College London in 2019 analyzed all the available evidence and estimated that 60.5 million Indigenous people had been living in the Americas in 1492 and that colonizers' arrival led to 56 million Indigenous deaths by 1600. The team calculated that more than 200,000 square miles of land (an area bigger than California, smaller than Texas) that had been cultivated to feed the larger population was abandoned and reclaimed by forest and jungle growth. "The regrowth soaked up enough carbon dioxide from the atmosphere to actually cool the planet," said *The Guardian*.

Compiling global climate data from 1000 to 1800, the scientists found that between 1577 and 1694, global temperatures dropped 0.15 degrees Celsius (0.27 Fahrenheit). This corresponded with a drop in carbon dioxide in air bubbles trapped in Antarctic ice cores from around 1570 to 1610. Plants absorb CO_2 from the atmosphere, and the denser growth of natural vegetation absorbs significantly more than fields planted with crops. The drop in CO_2 could not be accounted for by other natural phenomena, so the study concluded that the destruction of Indigenous communities, with their skilled agriculture labor, contributed to the drop in carbon dioxide and affected global temperatures. Points out climate science professor Ed Hawkins, "This new study demonstrates . . . that human activities affected the climate well before the industrial revolution began."

So, while we often focus on climate change and its impact on our atmosphere, we should also understand that it's been affecting the very ground we walk on, too, for a long time. Plants and soil play an active role in the absorption of carbon dioxide from the atmosphere, storing it for decades, centuries, even millennia. Beginning in the 1800s, colonization, forced labor of enslaved people, and more industrial farming practices, especially the use of the plow, increased the amount of land that was cleared and tilled. Tilling breaks up soil, bringing stored carbon dioxide to the surface, so much more CO_2 was driven into the atmosphere, and fewer trees were around to absorb it.

Industrial farming practices like plowing and planting the same crop in the same place every year have harmed the very soil meant to sustain us. Healthy soil is not simply dirt; it is a complex ecosystem teeming with life that recycles carbon and other nutrients back into the Earth, allowing plants and animals to grow. It holds water, filters potential pollutants, and provides a stable base for all of us. When the soil is not healthy—when the land is degraded, as scientists say—the effects are far-reaching and long-lasting. Degraded land doesn't sustain as many plants or produce as much food, and it can't absorb as much carbon, making climate change worse. Then, of course, climate change degrades the land even more. The United Nations recently confirmed what the early environmental conservationist George Washington Carver knew over a century ago: sustainable farming and soil cultivation is vital to our survival.

GEORGE WASHINGTON CARVER (1861?–1943) was a scientist and inventor born into slavery. He worked to encourage more sustainable farming practices by promoting growth of plants that could nurture and replenish our soil, like peanuts and sweet potatoes. "I love to think of nature as unlimited broadcasting stations, through which God speaks to us every day, every hour and every moment of our lives, if we will only tune in," Carver famously wrote in a 1930 essay, "How to Search for Truth."

Can't Take the Heat

Earth is now as warm as it was before the last ice age, 115,000 years ago, when the seas were more than twenty feet higher. In 1990, humankind emitted more than 20 metric gigatons—or 20 billion metric tons—of carbon dioxide. In 2018, we produced 36.6 metric gigatons—a record (one metric gigaton equals one billion metric tons, roughly twice the mass of all the people in the world). The global pandemic meant that things slowed down a bit, and CO_2 emissions fell by about 6.4 percent in 2020, largely due to a decline in vehicle usage in the United States. But emissions zoomed back up in 2021, which was roughly tied for the sixth-hottest year ever observed (with 2015–2021 marking the seven hottest years on record). "Without substantial collective action to curb

emissions, 2020 will register as little more than a blip in the global carbon record," cautioned one climate expert.

Research has shown that the rapid changes we're experiencing are due more to the increase in heat-trapping gases in our atmosphere than anything else. Those gases are increasing because of us. Nitrous oxide (nitrogen and oxygen) is part of the life cycle, but human activities like transportation, farming and industry, even wastewater treatment, are responsible for 40 percent of the gas's emissions. Methane (carbon and hydrogen) is emitted by wetlands and volcanic eruptions, but humans are responsible for more than half its emissions now, through such activities as raising cattle, fracking, and storing waste in landfills. Human activity has made carbon dioxide the biggest culprit of all. We produce more carbon dioxide than any other gas, mainly through the burning of fossil fuels in driving and powering electricity and industry, but also through deforestation and farming. In 2019, fossil fuels accounted for the vast majority of the world's energy consumption:

- Oil 33%
- Coal 27%
- Natural gas 24%
- Hydroelectric 6%
- Renewables 5%
- Nuclear energy 4%

The threat posed by global warming has been recognized by scientists for decades. Back in 2001, in a report on climate change requested by the White House, the National Academy of Sciences stated, "Greenhouse gases are accumulating in the Earth's atmosphere as a result of human activities, causing surface air temperatures and subsurface ocean temperatures to rise." Four years later, the academy joined the science academies of ten other nations in declaring "that the scientific understanding of climate change is now sufficiently clear to justify nations taking prompt action." Other leading U.S. scientific bodies, including the American Meteorological Society, the American Association for the Advancement of Science, and the American Geophysical Union, have since issued similar statements.

FRACKING: A process by which a pressurized liquid is injected into rocks underground as a way of extracting gas and oil. Fracking allows drilling companies to capture difficult-to-reach sources of oil and gas. Critics say that it releases toxic chemicals into the atmosphere, uses too much water, and keeps us hooked on fossil fuels.

The Intergovernmental Panel on Climate Change, an international group that assesses the science related to climate change, says it's clear that humans have influenced the climate. The increase in greenhouse gases due to human activity, the panel said in 2014, is "extremely likely to have been the dominant cause of the observed warming since the mid-20th century."

To put it simply: The more carbon dioxide in the atmosphere, the warmer the planet. And these days, there's a lot more.

In fact, more carbon has been released into the atmosphere since November 7, 1989 (the final day of one of the first international conferences about the dangers of global warming), than in the entire history of civilization preceding it.

Read that sentence again.

Yep, you read that right.

Nope, it doesn't make sense.

Today, we produce more carbon dioxide than our Earth can handle, and that is warming up the planet fast.

WHO'S "WE"? IMBALANCE AND INEQUALITY

Depending on what part of the world you live in, your carbon footprint is very different. Here, in metric tons of CO_2, is the average carbon footprint in 2018 of one person in each of these countries:

UNITED STATES 17.7	PHILIPPINES 1.5
CANADA 15.9	CAMEROON 0.4
RUSSIA 9.8	INDIA 1.7
CHINA 6.4	COLOMBIA 2.1

When you look at the total amount of carbon emissions produced in a country, you get a different picture. These countries were the top contributors of CO_2 to our atmosphere in 2018, in metric gigatons:

CHINA 9.95

UNITED STATES 5.42

INDIA 2.59

RUSSIA 1.69

JAPAN 1.14

Clearly some of us on this planet are doing a lot more damage than others.

What Does a Warmer Planet Mean?

So we've already heated things up more than one degree Celsius since the Industrial Revolution.

Let's imagine we go to two degrees Celsius. It would mean a long-term disaster. Unfortunately, long-term disaster is now the best-case scenario.

Three degrees warmer, and you've got a prescription for short-term disaster. We're talking stuff straight out of a Hollywood horror movie: Forests sprouting in the Arctic, the flooding and abandonment of most coastal cities, mass starvation.

If we keep going to four degrees: Europe—which in 2020 experienced its warmest year ever—will be in permanent drought. Vast areas of China, India, and Bangladesh will be claimed by desert.

Polynesia will be swallowed by the sea. The Colorado River will thin to a trickle. The prospect of a five-degree warming prompts some of the world's top climate scientists to warn of the fall of human civilization.

Yikes! doesn't even begin to cover it.

Continued global warming brings with it all kinds of risks:

- Mass migration from the regions most affected by climate change. Vulnerable people who have been denied access to the systems and resources needed to equip and protect themselves will feel the greatest impact.

- More destructive wildfires. By 2050, fires in the western United States are projected to burn twice as much land as they do now, with fire season starting weeks earlier and lasting longer. As temperatures rise

and conditions get drier, wildfires spread faster and are more difficult to put out. Higher year-round temperatures also increase the populations of insects that weaken and destroy trees.

- A loss of most if not all coral reefs on Earth. Even with only a 1.5-degree rise, we will lose 70–90 percent of our reefs; two degrees will kill off more than 99 percent of these dynamic and delicate ecosystems, which support 25 percent of marine life.
- More heat stress in cities. Over 350 million more people around the world would be exposed to deadly heat by 2050.
- A risk of damage to $1 trillion in public infrastructure and coastal real estate in the United States. Hundreds of billions of dollars in U.S. economic output will be lost.

We're seeing these things beginning to happen already.

Fire is a natural part of forest life in the western states, but California has seen a remarkable increase in very large wildfires in recent years. In 2020 alone, the state saw five separate fires burn more than 300,000 acres each. Before 2018 no single fire in the state had ever reached that size. The largest 2020 wildfire spread across a million acres. Fires not only remove vegetation that absorbs carbon dioxide, exposing soil to erosion, but burned soil doesn't absorb water as easily, which leads to floods and debris flows when the rains come. After a large fire in late 2017, torrential rain brought dangerous mudslides and dislodged boulders, killing twenty-three people and damaging more than four hundred homes.

Methods used to extract fossil fuels such as fracking and coal mining also destabilize the soil, resulting in sinkholes and landslides.

CLIMATE CHANGE MYTHS VS. FACTS

The climate has always changed, you said so yourself.

Yeah, but not this fast, remember? Nineteen of the twenty warmest years on record have all taken place after 2000.

Okay, but we still have plenty of cold weather, so . . . what's up with that?

Here's a good way to think about it, according to climate science professor Katharine Hayhoe: "Weather is like your mood, and climate is like your personality." Weather involves short-term changes in our atmosphere, while climate is about what happens over long periods. Climate change does cause the Earth's average surface temperature to rise over time, but it also causes more storms, flooding, and drought.

There's no definite proof of climate change; a lot of scientists are still debating the issue.

Don't believe the hype. More than 97 percent of climate scientists agree that global warming is happening and that we humans are the cause of it.

Chapter 3
NO JUSTICE, NO PEACE

Climate Injustice

Often when we think of refugees, we think of vulnerable people trying to escape political conflict, wartime violence, and deadly threats from other human beings.

But the Red Cross currently estimates that more refugees flee environmental crises than violent conflicts. Starvation, drought, coastal flooding, and desert expansion will force hundreds of millions of people to run for their lives. Mass migration will lead to instability across the world, increasing conflicts and, ultimately, war.

And even though there is no escape from the effects of climate change for any of us, some places will experience the negative effects harder and sooner. How do we know? Because it's already happening.

[Environmental justice is] the principle that all people are entitled to equal environmental protection regardless of race, color or national origin. It's the right to live and work and play in a clean environment.
—**Robert Bullard**, environmental justice advocate

Let's take another look at that Industrial Revolution. By moving from horse, water, and wind power to engines and other machinery steam-powered by burning coal and wood, the United States and countries in Western Europe got an early start on mass manufacturing and the fossil-fuel burning that's been pumping extra carbon dioxide into our air for over two centuries.

The Western world has been going full blast ever since, building wealth and technological advances and the modern conveniences that many in these countries enjoy. But those benefits come at a high price—and that price is paid by everyone.

"The rich got rich on high-carbon growth," as one climate economist has said, "and it's the poor people of the world—whether they be poor people in rich countries or poor people in poor countries—who suffer earliest and most."

For example, climate change will have the greatest effect on agriculture, a primary source of income in the world's poorest countries and a key part of their economy. In countries rich and poor alike, crop losses will lead to rising food costs, which will affect the poorest first, no matter where they are. As food costs rise, poor people will be forced to spend more of their income on food. The strain could lead to malnutrition, even starvation.

Other health risks come with climate change. Higher temperatures will expose millions more people around the world to malaria and other diseases carried by insects and parasites. More frequent extreme weather events, like floods and droughts, will contaminate more drinking water. Diarrhea will be more common, leading to thousands more deaths among children. All these health risks will have a greater impact on people with lower incomes, who have less access to health care and pay more out of pocket for it.

There is a long history of unequal impact when it comes to changes in the environment. For instance, science in the early 1970s showed that man-made chemicals known as PCBs (polychlorinated biphenyls) were harmful to the environment and humans. That led to a ban on U.S. production, and laws regulating proper PCB disposal.

But in North Carolina, one company devised a plan to dispose of PCB-contaminated soil by dumping it along rural roadways. In the summer of 1978, it contaminated over two hundred miles of highway shoulders in fourteen counties, and the state was left to clean it up. What did the state decide to do?

It decided that rural Warren County, mostly Black and low-income, was the perfect place to dump all that toxic dirt.

The residents fought back vigorously. They showed up at public hearings to express their outrage at the proposal.

"If it means we have to stand bodily in front of bulldozers, trucks, and moving equipment, to give our very lives to save lives in the future," a local pastor said, "I say it is our right and our duty to sacrifice ourselves."

And that's exactly what they did. When lawsuits failed to stop what the governor called "the Cadillac of landfills," residents called in civil rights activists, and together they lay down in the streets, trying to protect the community with their lives. Hundreds were

arrested during the weeks of protests, including another pastor, Reverend Benjamin Chavis, with the United Church of Christ's Commission on Racial Justice. Fifty thousand tons of toxic dirt was dumped anyway.

Calling the state's actions "environmental racism," Chavis led the commission in its landmark 1987 study, *Toxic Wastes and Race in the United States*. The report found that race was the strongest factor in predicting the location of hazardous waste facilities in the United States, "more powerful than household income, the value of homes, and the estimated amount of hazardous waste generated by industry."

Money makes a difference. Global powerhouses like the United States have more resources to protect them from the more dire effects of climate change. As activist Rajit Iftikhar, a software engineer at Amazon, puts it, "Americans continually underestimate how bad the climate crisis will be for everyone in years to come because it's easy to ignore those who are facing the crisis right now."

Poor people are already being forced from their homes and lives because of problems that emerge from climate catastrophes. It's a situation that makes bad things like political and financial uncertainty much worse.

For instance, the three-year drought that began in Syria in 2007 led to crop failure and hunger, driving hundreds of thousands of people into the cities, worsening social and economic tensions that led to protests and civil war. Typhoons and storms in the Pacific have displaced millions of people in the Philippines.

Hurricanes like Katrina in 2005 and Harvey and Maria in 2017 have devastated cities and communities in the United States, but "climate refugees" usually come from regions that are not responsible for the energy consumption that's causing the most damage. They don't have the resources to respond to a problem that they didn't create. As the situation deteriorates, the world's most impoverished people will be hit the hardest. Black and Brown people will disproportionately suffer from natural disasters, fertile land deterioration, food and water shortages, and migratory chaos. Climate change amplifies social inequity. It disadvantages the disadvantaged, oppresses the oppressed, discriminates against the discriminated against. "This is a conversation that many people don't want to have," says activist and author Vanessa Nakate. "You know, people don't like mixing climate and, for example, race, or climate and gender.

And in the end, the destructive forces that humans have unleashed will affect us all.

So, this sounds really, really bad. We've got a problem. We know we've got a problem. Climate change is serious. Actually, wait—let's be real. This is not about climate "change." This is about climate crisis. In talking about this problem, the very words we use matter. In 2019, London's *Guardian* newspaper made an official decision to change the language used in its coverage of climate issues. Instead of *climate change*, the preferred terms are now *climate emergency*, *crisis*, or *breakdown*, and *global heating* is favored over *global warming*. The logic behind the change is that the phrases *climate change* and *global warming* sound pretty gentle when scientists are talking about a looming catastrophe for all of humanity.

So we've got to figure out a solution, right?

The good news is that we already have!

Thirty years ago, scientists concluded that there was a way to avoid disaster. They did the research, they shared their findings, and a lot of powerful people were concerned, even alarmed, and prepared to act. Many not-so-powerful people agreed, because they had known a lot of this all along, and lived accordingly. And all signs pointed to this: The United States, considered by many to be the most powerful country in the world, was not only doing the most damage, but it had the resources to lead the way to the most comprehensive solution.

And there was one decade in particular, from 1979 to 1989, when the greenhouse effect became widely known, when every-day people began to understand the threat posed by climate change, when the United States had an excellent chance, a golden opportunity, to solve this problem, to do the right thing . . . and didn't.

Not only that, the country got on the fast track to self-destruction.

Right now, today, as you read this, we still have time before those worst-case scenarios are our reality. But that's the thing. We've always thought we had time. And how did our governments and systems use that time? Well . . . we've already had decades—decades increasingly punctuated by climate-related disasters like hurricanes, fires, heat waves, floods, and more—and we've done nearly everything possible to make the problem worse.

That's right: We have used that time to actively make things worse. It seems irrational.

Yep, the United States had a chance to save the world—to actu-ally be those superheroes in books and movies and our dreams—and this country put it off. Why? That's not who we are, right? But . . . that's what we did, and what we did says a lot about who we are . . . and maybe what we'll do next.

"A Race without the knowledge of its history is like a tree without roots," wrote Black historian Charles Seifert in 1938.

So let's look back at who we were

and dig up the roots of who we are,

and then maybe we'll give ourselves the opportunity to become who we want to be.

Part II

That we came so close, as a civilization, to breaking our suicide pact with fossil fuels can be credited to the efforts of a handful of people—scientists from more than a dozen disciplines, political appointees, members of Congress, economists, philosophers, and anonymous bureaucrats. They were led by a hyperkinetic lobbyist named Rafe Pomerance and a guileless atmospheric physicist named James Hansen who, at severe personal cost, tried to warn humanity of what was coming. They risked their careers in a painful, escalating campaign to solve the problem, first in scientific reports, later through conventional avenues of political persuasion, and finally with a strategy of public shaming.

Their efforts were shrewd, passionate, robust.

And they failed.

What follows is their story, and ours.

—**Nathaniel Rich,** *Losing Earth: A Recent History*

Chapter 4
A HISTORY OF PROTEST

Sisters Doing It for Themselves

In many parts of the world, particularly those most vulnerable to climate change, groups of people have long nurtured a sense of respect, even reverence, for the natural world. And they have stood up to threats.

In 1730, over 350 villagers in northwest India gave their lives to protest the cutting down of trees they considered sacred. They were members of the Bishnoi faith community, a Hindu sect founded centuries earlier on principles of conservation and environmental stewardship.

The maharajah, the ruler of Jodhpur, had ordered the trees be cut to build a new palace. A woman named Amrita Devi pleaded

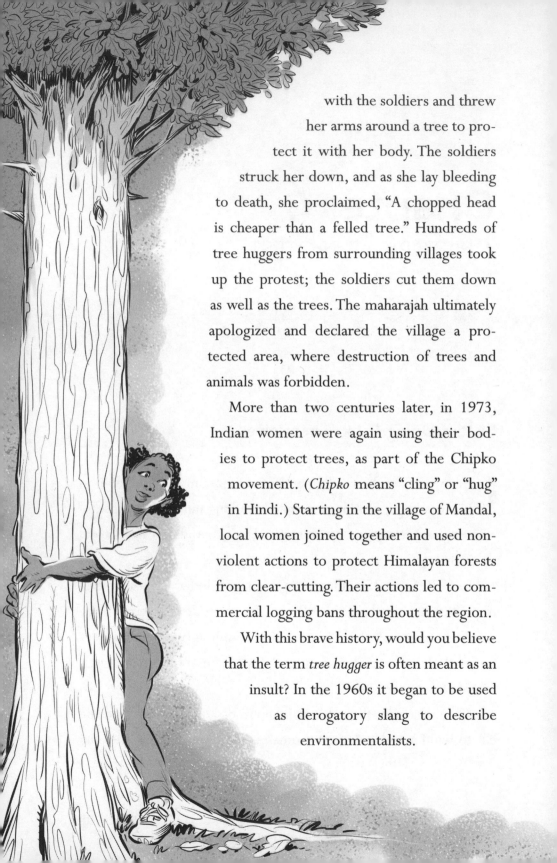

with the soldiers and threw her arms around a tree to protect it with her body. The soldiers struck her down, and as she lay bleeding to death, she proclaimed, "A chopped head is cheaper than a felled tree." Hundreds of tree huggers from surrounding villages took up the protest; the soldiers cut them down as well as the trees. The maharajah ultimately apologized and declared the village a protected area, where destruction of trees and animals was forbidden.

More than two centuries later, in 1973, Indian women were again using their bodies to protect trees, as part of the Chipko movement. (*Chipko* means "cling" or "hug" in Hindi.) Starting in the village of Mandal, local women joined together and used non-violent actions to protect Himalayan forests from clear-cutting. Their actions led to commercial logging bans throughout the region.

With this brave history, would you believe that the term *tree hugger* is often meant as an insult? In the 1960s it began to be used as derogatory slang to describe environmentalists.

DEFORESTATION AND YOUR DINNER

Forests of trees, like all green plants, absorb carbon dioxide as part of their growing process. When farmers and loggers cut down or burn trees—to harvest the wood or clear space for crops and livestock grazing—the carbon stored in the trees is released into the air as carbon dioxide. Ten percent of all greenhouse gas emissions come from the destruction of tropical forests, mainly for the production of globally traded beef, soybeans, palm oil, and wood. Slowing deforestation by supporting sustainably produced meat and agricultural products is one way you and your family can fight global warming at the dinner table.

Many women throughout history have been willing to lead the fight for responsible environmental stewardship, trying to improve how we use and care for our natural world.

Wangari Maathai, for example, mobilized the women of Kenya to plant more than thirty million trees and inspired a United Nations campaign that planted millions more.

Born in Nyeri in 1940, young Wangari Maathai collected firewood for her mother, who told her not to touch the fig trees. "That is a tree of God. We don't cut it. We don't burn it. We don't use it," her mother said. "They live for as long as they can, and they fall on their own when they are too old."

Yet as she got older, Maathai saw Kenyan forests cleared, replaced with commercial plantations and development. By the 1970s, rural women were reporting that their streams were drying up and their harvests were poor.

Turns out, the loss of trees was a key part of the problem. Tree roots "are able to go down into the underground rock," Maathai later explained. "They are able to break the rock, and they are able to bring some of the subterranean water system up nearer to the surface, and so they were responsible for many of the streams that dotted the landscape. In many ways, therefore, they were part of the water system in the area, and so they served a very important purpose." Wangari Maathai had gone on to become the first woman from that region to earn an advanced academic degree. After studying science in America and Germany, Dr. Maathai believed that she returned home with even more than a fancy degree—she had a mission. She believed that the civil rights movement in the United States in the 1960s greatly influenced her sense of justice.

Remembering her mother's words, Maathai realized that many traditional practices come from people in the past "having learned how to live within the environment in which they found themselves."

Thinking about the impact of environmental destruction on Indigenous communities, particularly women, Maathai began the Green Belt Movement in 1977 to promote tree-planting as a way to improve the land as well as the livelihoods of people. She soon understood, however, that sustainable land use could not take place without good government. So the Green Belt Movement empowered the people to rally against the corruption that leads to environmental mismanagement. The Kenyan government had her jailed and beaten for her efforts.

The Nobel Committee awarded her the Peace Prize in 2004. As the committee noted, "Peace on earth depends on our ability to secure our living environment."

THE GREAT GREEN WALL

The spirit of the Bishnoi tree huggers and Wangari Maathai can be seen in projects like the Great Green Wall, a plan to restore land in Africa and prevent it from becoming desert. Launched in 2007 by the African Union, this project in the Sahel savannah south of the Sahara involves not only planting trees and vegetation but also techniques that Indigenous people have learned over time, such as protecting saplings that naturally sprout up on farmland. The ultimate goal is healthy soil that will provide food, jobs, and climate resilience in one of the world's poorest regions. In 2021, an international coalition invested an additional $14 billion to accelerate the project.

Around the world, in small communities and countries already suffering the environmental cost of the Industrial Revolution, people have been paying attention to the natural world and using ancient wisdom to protect the land and fight for environmental justice.

In the United States, which has a long history of ignoring the wisdom of the Indigenous people it tried to remove, the cry for environmental protection and justice has come from the research and activism of dedicated individuals.

Rachel Carson documented the environmental and health effects of pesticides in her landmark 1962 book, *Silent Spring*. It galvanized the public and helped spark the environmental movement.

Senator Gaylord Nelson of Wisconsin was the creator of Earth Day. His idea was to give the environmental movement the same visibility as the civil rights and anti–Vietnam War movements. On April 22, 1970, twenty million people from all walks of life and political parties came out across the United States to celebrate the Earth in a campaign to promote awareness of air and water pollution.

The emergence of a national environmental movement helped to bring about important legislation to protect the environment and human health. President Richard Nixon created the Environmental Protection Agency and signed the Clean Air, Clean Water, and Endangered Species Acts into law. People learned how their daily actions affect the very world we live in.

Yet on that first Earth Day in 1970, the senator from Wisconsin knew that the issues ran deeper. "Environment is all of America and its problems," Nelson said. "It is a hungry child in a land of affluence. It is housing that is not worthy of the name; neighborhoods not fit to inhabit."

So, how would the United States, a major international power, proceed? What would the next steps on its own soil be? If we dug a little deeper, what would we find beneath the surface?

Digging Deeper

In the 1970s, digging deeper into U.S. soil brought up more than worms, and the work of Robert D. Bullard illuminated that.

Today Bullard is considered the father of environmental justice, but back in 1979 he was a young sociology professor collecting data for his lawyer wife's case. He was drafted into the environmental movement, as he says, and like his inspiration, the famous sociologist W. E. B. DuBois, he became an agent of change.

His wife, Linda McKeever Bullard, was representing Black residents of Houston who were fighting to keep a landfill out of their community. In the landmark lawsuit, she argued that the decision by the Texas Department of Health to allow the landfill was motivated by racial discrimination. Her case was the first to use civil rights law to challenge the siting of a waste facility.

Bullard served as an expert witness. He found that 100 percent of the city-owned landfills in Houston were in Black neighborhoods, along with three-fourths of privately owned landfills and inciner-ators. In a city without zoning laws, someone had to be making those decisions, so he did some more research.

"I got hooked. I started connecting the dots in terms of hous-ing, residential patterns, patterns of land use, where highways go . . . and how economic-development decisions are made," he said. "Without a doubt, it was a form of apartheid where whites were making decisions and black people and brown people and people of color, including Native Americans on reservations, had no seat at the table."

Bullard's research was the first documentation of environmen-tal discrimination.

REDLINING

During the Great Depression, a federal agency began "grading" neighborhoods on their lending risk. Anyone not of northern European descent was considered a risk to property values, and the agency literally drew red lines around Black and ethnic minority neighborhoods on the map and labeled them "hazardous." Loans in redlined neighborhoods were expensive and hard to come by, making it difficult for residents to buy homes and accumulate wealth. Redlining was banned in 1968, but its effects persist. A 2018 study showed that three out of four of those redlined neighborhoods struggle economically today.

He went on to study discriminatory environmental practices and policies across the Southern states, exploring how the civil rights movement and the fight for just environmental practices were linked. The Reverend Dr. Martin Luther King Jr., for example, joined sanitation workers striking in Memphis, Tennessee, to protest unfair pay and unsafe conditions in 1968. While in Memphis, King was assassinated.

Bullard became one of a group of scientists, activists, and Black leaders who advocated for marginalized communities on environmental issues. Their work pushed the EPA to look at the evidence that minority and low-income communities were bearing a higher burden of environmental risk. Finding the evidence clear, the EPA established the Office of Environmental Justice in 1992.

Bullard was also one of the leaders of the first National People of Color Environmental Leadership Summit in 1991, where delegates gathered to build a movement "to fight the destruction and taking of our lands and communities." In adopting seventeen Principles of Environmental Justice, they pledged "to respect and celebrate each of our cultures, languages and beliefs about the natural world," to promote "environmentally safe livelihoods," and to secure "our political, economic and cultural liberation." The summit "was a turning point in the environmental movement," writes historian Jeff Chang. A coalition of frequently marginalized voices, the summit included more than 1,000 attendees from the fifty states and Puerto Rico, Chile, Mexico, and the Marshall Islands, and featured experts like Black environmental justice activist Hazel Johnson; Wilma Mankiller, the first woman to serve as Principal

Chief of the Cherokee Nation; and Pam Tau Lee, cofounder of the Asian Pacific Environmental Network. The historic four-day event, continues Chang, "made it clear that BIPOC communities are at the frontlines of protecting our land, air, and water."

It's impossible to separate issues of environmental justice and economics from race, Bullard believes, even if it's hard to discuss. "People get uncomfortable when questions of poor people and race are raised." Nevertheless, this author of eighteen books continues to raise these questions with people in power. In 2021 President Joe Biden named him to his new White House Environmental Justice Advisory Council.

Bullard sees climate change as the leading problem of our day. "We sometimes forget that climate change is much more than simply parts per million [of greenhouse gas emissions]. It is an equity issue. It affects some people directly." In 2011 he cofounded the HBCU Climate Consortium to help train the next generation of climate leaders at historically Black colleges and universities—because climate change, like hazardous waste disposal, disproportionately harms people of color and poor people.

Significant and Damaging

Not long after Bullard's discoveries beneath the dirt in the South, a young man named Rafe Pomerance learned a few equally shocking things about what was going on up in our atmosphere.

Pomerance was the deputy legislative director of Friends of the Earth, an environmental advocacy organization in Washington, DC.

In the spring of 1979, he sat in his stuffy, windowless office reading a tedious government report. It was a long document with a dull title—*Environmental Assessment of Coal Liquefaction*—stuffed with technical language and boring details. It wouldn't be a page-turner for many people, but ensuring clean air was Pomerance's job—so he kept reading. When he got to page 66—and he might have been the first person to make it to page 66 of this particular report—he came across something shocking:

> *A report by the National Academy of Sciences ... warns that continued use of fossil fuels as a primary energy source for more than 20 to 30 more years could result in increased atmospheric levels of carbon dioxide. The greenhouse effect and associate global temperature increase could ... be both "significant and damaging."*

Wait, what?

Pomerance read it again, just to be sure. It didn't make sense. He wasn't a scientist (he'd studied history in college), but in his role at Friends of the Earth, he was one of the most connected environmental activists in the country. He'd been immersed in these issues for years. He figured he knew everything there was to know about air pollution. He wondered, What was this "greenhouse effect"? And if the burning of coal, oil, and natural gas could bring about a worldwide catastrophe, why hadn't anybody told him about it?

MORE ABOUT THE GREENHOUSE EFFECT

The greenhouse effect refers to the warming trend of the Earth's temperatures, now happening at an increasingly faster pace. NASA's research of the atmospheres of Venus and Mars helped scientists understand what could happen if we let greenhouse gases get way out of control. This warming trend won't cause just a little perspiration or a minor inconvenience. It's a dangerous cycle—as the Earth heats up, more water becomes water vapor in our atmosphere. That means more heat is trapped, which means more water vapor, and on and on. It doesn't sound good at all, but it's called a "positive feedback loop" because it speeds up the rise in temperatures.

He showed the unsettling paragraph to his office mate. Had she ever heard of the greenhouse effect? Was it really possible that human beings were overheating the planet?

His office mate shrugged. She hadn't heard about it either.

Chapter 5
SOUNDING THE ALARM

Talking to Some Superheroes

After hearing about the greenhouse effect, Rafe Pomerance learned that a group called JASON had studied what would happen if carbon dioxide in the atmosphere doubled. The members of the group, known unofficially as the Jasons, were like a scientist version of a team of superheroes who join forces in times of galactic crises.

This group of elite scientists and engineers were summoned by the U.S. government to solve major global problems—to save the world—by using science and technology. In 1978, the Jasons concluded that human civilization could contribute as much carbon dioxide in the atmosphere by 2035 as the planet had done in the previous 4.6 billion years. If nothing changed, large areas of the United States would suffer prolonged droughts, and the oceans would rise by sixteen feet. The Jasons' report was delivered to scientists around the world and senior officials across the U.S. government.

Pomerance needed to learn more. Maybe the Jasons could answer all the questions he had about this mysterious greenhouse effect he'd read about. He asked for a meeting with Gordon MacDonald, the lead author of the JASON report.

"I'm glad you're interested in this," MacDonald said the first time they got together.

"How could I not be?" said Pomerance. "How could anyone not be?"

MacDonald spoke for two hours. As he traced the history of humanity's understanding of the problem, explaining the fundamental science, Rafe Pomerance got increasingly fired up. He proposed that the two of them explain this to the people in power in the government, all the way up to the White House.

Pomerance figured that if it was confirmed that human civilization was working on a fast track to its own extinction, the president would be forced to act.

The Bad News Gets Worse

In 1979, James Hansen figured he was the only NASA scientist who had not dreamed of outer space as a child. Hansen dreamed only of baseball. He loved statistics; he was interested in numbers, and the stories they told. As he grew older, and began to study distant worlds, he learned that the numbers measuring the composition of their atmospheres revealed fascinating stories about each planet's past and future. It made him curious about the Earth's atmosphere. He wanted to understand what future its own numbers might predict.

At NASA, Hansen had studied Venus to understand why its surface was so hot. Around the same time that Rafe Pomerance was learning about global warming, Hansen began to apply the lessons

he had learned from Venus to the Earth. The answers alarmed him.

The Jasons' conclusions, Hansen calculated, were wrong.

The situation was . . . WORSE.

When the president's top science adviser asked a group of top experts from the National Academy of Sciences to review the Jasons' warnings, Hansen was invited to share his discoveries. The 1979 publication of the academy's final report, *Carbon Dioxide and Climate: A Scientific Assessment*, was not accompanied by big fanfare, a parade, a banquet, or even a press conference. But within the highest levels of the federal government, the scientific community, and the oil and gas industry (in other words, within the group of people who had begun to concern themselves with the future habitability of the planet), the report settled things.

The academy's study group had considered everything known about the subject and had distilled it to a single number: three.

THREE DEGREES CELSIUS: A LITTLE NUMBER THAT MEANS A LOT

If carbon dioxide in the Earth's atmosphere doubled in 2035 or thereabouts, global temperatures would increase between 1.5 and 4.5 degrees Celsius, with the most likely outcome falling in the middle: a warming of three degrees. The last time the world was three degrees warmer was during the Pliocene epoch, three million years ago, when beech trees grew in Antarctica, the seas were eighty feet higher, and wild horses galloped along the Canadian coast of the Arctic Ocean.

Now James Hansen had questions. A warming of three degrees would be nightmarish, but unless carbon emissions ceased suddenly, three degrees would be only the beginning. The real question was whether the warming trend could be reversed. Was there time to act?

The report warned, "A wait-and-see policy may mean waiting until it is too late." But how would a global commitment to cease burning fossil fuels come about, exactly? Who had the power to make such a thing happen?

Hansen had to find out.

Avoiding Disaster

Rafe Pomerance didn't need any more government reports or expert consultations: He already understood that the burning of fossil fuels had to end in order to avoid global disaster. He believed that a solution was technically possible. But he wondered what it would take to force humanity to act. Even if everybody understood the dangers of global warming, would we, as a society, be willing to take action to prevent it?

It sounds like an easy question to answer, right? Of course, if we see that we're destroying the Earth, that we're on the fast track to wipe out civilization, we're gonna stop, right? Surely, we'll take a step back and say, *Whoa*.

We are facing a period when society must make decisions on a planetary scale . . . Unless the peoples of the world can begin to understand the immense and long-term consequences of what appear to be small, immediate choices, the whole planet may become endangered.
—**Margaret Mead** (1901–1978), anthropologist

But Rafe Pomerance understood that it was . . . complicated.

Back in the mid-1700s when we started burning fossil fuels in a major way, things quickly got out of control. As the use of coal accelerated in the 1800s, cities grew more crowded and congested with factory workers—and factory smoke. In English textile towns like Leeds and Manchester, campaigns against the smoke sprang up as early as 1842. Still, by 1879 chronic bronchitis, caused by inhaling smoke, was Britain's biggest killer. Did we stop and say *whoa* then?

In the twentieth century, the explosive growth in gasoline-burning engines in our cars and trucks only amplified the air pollution created by coal-fired factories and power plants. In 1943, there was so much smog in Los Angeles that residents believed the city to be under attack from the Japanese. The 1952 Great Smog of London, mostly caused by coal burning, brought such a heavy blanket of poisonous smog over the city for four days that thousands of people died.

Fortunately, clean air laws removed much of the smoke from our air. But unfortunately, as our air became clearer and technology grew more sophisticated, we thought less of the damage we were still doing. Even our most routine activities—driving in a car, turning on a light switch, using an air conditioner—required combustion of vast quantities of fossil fuels. Even if we understood this, we never seemed to think about it. Whether or not we thought about it, the problem was getting worse by the day. Pomerance knew we were living on borrowed time, and the bill would come.

An earlier government report had warned that humanity's fossil fuel habit would lead to a host of "intolerable" and "irreversible" disasters, but it categorized the best available remedy—a transition to renewable energy—as far-fetched. "Any government action requires political consensus," concluded the authors. "Such consensus may be difficult to achieve."

Difficult indeed. And still . . . the question remained. When the threat seemed so distant, would we change? We say we worry about our children's futures, our grandchildren's. But . . . how much? And what about the great-grands? The great-greats? How much do we really care about the future? According to the scientists at the time, the amount of change needed would mean some public sacrifice. Would people leave the personal comfort of their cars for crowded public buses? Hang their clothing on racks instead of using electric- or gas-powered dryers? How much would we have to do?

RENEWABLE ENERGY: Energy that comes from sources like wind, water, sunlight, and biomass (like wood, natural waste, corn-based fuel), which don't run out because they're naturally replenished over time. It's also called "clean" energy because it doesn't produce those nasty greenhouse gases that we've long known are not cool at all.

A lot, the scientists told the people in power.

And told and told.

They told the fossil fuel industry, who already knew. They'd been studying the problem since the 1950s.

They told the politicians, who were focused on problems that could be solved in just a few years, just in time to help them win the next election.

Some people simply didn't believe the scientists. Some didn't want to believe them.

Chapter 6
ON THE BRINK

Warning Signs

After the publication of the National Academy of Science's 1979 report, the gas and oil company Exxon decided to create its own carbon dioxide research program. Exxon wasn't that focused on how much the world would warm. It wanted to know how much of the warming could be blamed on Exxon.

Exxon had been tracking the carbon dioxide problem since before it was Exxon. In 1957, scientists from its predecessor, Humble Oil, published a study analyzing "the enormous quantity of carbon dioxide" contributed to the atmosphere since the Industrial Revolution "from the combustion of fossil fuels." Even then, the notion that burning fossil fuels had increased the concentration of carbon in the atmosphere went unquestioned by Humble's scientists.

In 1957, a scientist who had led the development of the hydrogen bomb told members of the American Chemical Society, which included engineers from oil and gas companies, that the exploitation of fossil fuels might bring about climate change; he said it again two years later at a centennial celebration of the American oil industry in New York City. "When the temperature does rise by a few degrees over the whole globe," he told the assembled dignitaries, "there is a possibility that the icecaps will start melting and the level of the oceans will begin to rise."

There was warning after warning.

The ritual repeated itself every few years.

Industry scientists reviewed the problem, finding good reasons for alarm and better excuses to do nothing. Why should they, when almost nobody seemed worried, not the U.S. government nor even the environmental movement. As a 1972 report for the Department of the Interior put it, climatic changes would probably not be apparent "until at least the turn of the century." The danger seemed far away—it was easy to put off worrying. And energy use had always meant economic growth—the more fossil fuels we burned, the better our lives became. Why mess with that?

But time was running out. On April 3, 1980, a Massachusetts senator held the first congressional hearing on carbon dioxide buildup in the atmosphere. Three months later, President Jimmy Carter signed the Energy Security Act of 1980, which directed the National Academy of Sciences to start a multiyear, comprehensive study, to be detailed in a report called *Changing Climate*, that would analyze the social and economic consequences of climate change.

A CONCRETE PICTURE

If cement were a country, it would be the third-highest emitter of carbon dioxide, after China and the United States. Cement is the key ingredient in concrete, the world's most popular building material, and the production of cement accounts for roughly 8 percent of global carbon dioxide emissions. Cement is made by heating limestone in high-temperature kilns that burn huge amounts of fossil fuel. Limestone is rock made of the remains of sea creatures, so it contains a lot of carbon in itself. Heated in kilns, limestone releases that carbon as CO_2. About two-thirds of cement's carbon dioxide emissions come from the limestone; the rest comes from the fossil fuels used to heat the kilns and transport the material.

Most urgently, the National Commission on Air Quality, at the request of Congress, invited two dozen experts to a meeting in Florida to develop climate legislation.

It appeared that some decree from the federal government to restrict carbon emissions was inevitable. The warnings from the Jasons had caused alarm, and the confirmation from the National Academy of Sciences demanded serious action.

All Talk

So the scientists, activists, politicians, businesspeople, and other interested parties got together. Rafe Pomerance was there. They had three days to figure this whole carbon thing out. There was no formal agenda, just a young moderator from the government and a few pamphlets—the reports from JASON and the National Academy of Sciences—on each seat. The meeting was held at a Florida beachfront hotel, and the experts looked out at white sands and tranquil waters as they began their discussion.

"Would anyone like to break the ice?" the moderator asked, failing to recognize his own pun.

"We might start out with an emotional question," proposed an economist. "The question is fundamental to being a human being: Do we care?"

The response was swift and strong. Of COURSE they cared!

"In caring or not caring," said an engineer, "I would think the main thing is the timing." In other words, it was not an emotional question, but an economic one: How much did we value the future?

One scientist warned that people had a tendency to leave problems to the "eleventh hour, the fifty-ninth minute." (Sound familiar?)

There was general agreement that some kind of international agreement would be needed to keep atmospheric carbon dioxide at a safe level. But . . . nobody could agree on what that level was. And any policy that restricted the use of energy would cause trouble.

Another scientist pointed out some hard facts:

- If the whole country stopped burning carbon that year, it would delay the arrival of the doubling threshold by only five years.
- If the entire Western world somehow managed to stabilize greenhouse

gas emissions, it would forestall the inevitable by eight years.

- The only way to avoid the worst was to stop burning coal. Yet China, the Soviet Union, and the United States, by far the world's three largest coal producers, were burning coal at a rapidly accelerating rate.

What if, suggested a politician, they thought about this as a solution instead of a problem? Even if the coal and oil industries collapsed, new renewable energy technologies, like solar, would thrive, and the broader economy would be healthier for it.

Whoa, whoa, whoa! A scientist for the gas and oil industry didn't like that kind of talk. "We are not going to stop burning fossil fuels and start looking toward solar or nuclear fusion and so on. We are going to have a very orderly transition from fossil fuels to renewable energy sources."

Pomerance wanted to debate this point, but he didn't get the chance. It was time for lunch. Lunch, the other members of the group decided, was more important.

After lunch, they talked. And talked. And talked some more. Rafe Pomerance got more and more worried and frustrated. Why not propose a new national energy plan? "There is no single action that is going to solve the problem," he said. "You can't keep saying, 'That isn't going to do it,' and 'This isn't going to do it,' because then we end up doing nothing."

The talk continued.

Some wondered whether it was time to classify carbon dioxide as a pollutant under the Clean Air Act and regulate it as such. This was received like a belch.

Others wondered, Did the science really support such an extreme measure?

Yes, said Pomerance, the reports from scientists did exactly that. He was beginning to lose his patience, his civility, his stamina. "Now, if everybody wants to sit around and wait until the world warms up more than it has warmed up since there have been humans around—fine. But I would like to have a shot at avoiding it."

They needed to draft some policy proposals the next day. But when the group reconvened after breakfast, they immediately became stuck on a sentence in their introductory paragraph declaring that climatic changes were "likely to occur."

"Will occur," proposed one.

"What about the words *highly likely to occur*?" asked another.

"Almost sure."

"Almost surely."

"Perhaps you can use *will occur*," said someone, "and quantify the changes."

"Changes of an undetermined——"

"Changes as yet of a little-understood nature?"

"Highly or extremely likely to occur."

"Almost surely to occur?"

These two dozen experts, who shared the same scientific understanding and had made a commitment to Congress, could not draft a single paragraph. They never got to policy proposals. They never got to the second paragraph. The final statement was signed by only the moderator, who phrased it more weakly than the declaration calling for the workshop in the first place.

Rafe Pomerance had seen enough.

The United States wouldn't act unless a strong leader persuaded it to do so, someone who could speak with authority about the science, demand action from those in power, and hazard everything in pursuit of justice. Pomerance knew he wasn't that person; he was an organizer, a strategizer . . . His job was to build a movement. And he knew that every movement, even one backed by widespread consensus, needed a hero. He just had to find one.

The Situation Heats Up

Four days after that meeting, Ronald Reagan was elected president of the United States. He had promised to decrease the size of the

federal government and to cut back on federal rules and regulations. Rafe Pomerance wondered whether what he thought was the beginning of a solution was in fact the end of one.

In the following months, Reagan floated plans to close the Energy Department, increase coal production on federal land, and deregulate surface coal mining. He almost shut down the EPA, and ended up doing the next best thing—hiring a boss who cut the agency's budget and staff by a quarter.

In the midst of this carnage, the president's own environmental advisers submitted a report warning that fossil fuels could "permanently and disastrously" alter the global atmosphere, leading to "a warming of the Earth, possibly with very serious effects." It urged the government to give high priority to the greenhouse effect in national energy policy and to establish a maximum level of carbon dioxide in the global atmosphere.

President Reagan's response? *Nah*. He declined to act on his own advisers' warning. Instead he considered firing them.

Meanwhile, the general public was beginning to develop a vague understanding of the issue, thanks to national headlines proclaiming apocalypse: "Another Warning on 'Greenhouse Effect,'" "Global Warming Trend 'Beyond Human Experience,'" "Warming Trend Could 'Pit Nation Against Nation.'" Even *People* magazine got into the act. Back when Pomerance and Gordon MacDonald were explaining the issues to the people in power, the magazine published a photo of MacDonald standing on the Capitol steps, pointing above his head to where the water would reach when the polar ice caps melted. "If Gordon MacDonald is wrong, they'll laugh," the article began. "Otherwise, they'll gurgle."

But Rafe Pomerance knew that to generate major public attention, you needed major events. Studies were fine; speeches were good; news conferences were better. Congressional hearings, however, were best. Those had drama—witnesses sipping nervously from their water glasses, the audience transfixed in the gallery. There were villains. There was tension. There was a story! You couldn't hold a congressional hearing without a scandal, however, or at least a scientific breakthrough.

And then he had one. On August 22, 1981, the *New York Times* ran a front page story, "Study Finds Warming Trend That Could Raise Sea Levels." NASA scientists had found that the world had already begun to warm in the past century. What's more, they calculated that the warming trend would continue, faster than before. Most unusual of all, the study ended with a policy recommendation: In the coming decades human civilization should develop alternative sources of energy; fossil fuels should be used only "as necessary." The lead author was James Hansen.

Pomerance needed to meet this guy. Could Hansen be the hero Pomerance was looking for?

Hansen invited Pomerance to his stuffy, cramped office in New York City. In order to get to Hansen's desk, Pomerance had to walk carefully around teetering stacks of papers—Hansen's studies of Venus, Mars, and Earth.

As Hansen spoke, Pomerance listened and watched. He liked what he heard and saw; Hansen seemed like the type of guy who would perform well before Congress: trustworthy, brilliant, serious, tough. And Hansen had his own reasons to go to Washington—the Reagan administration wanted to cut his research budget, and he had to convince them that this was a bad idea. Pomerance quickly saw that unlike most scientists in the field, Hansen was not afraid to explain how his research could lead to political action. He was perfect.

"What you have to say needs to be heard," said Pomerance. "Are you willing to be a witness?"

Chapter 7
FROM REACTION TO NO ACTION

Make Some Noise

A few members of Congress were already talking about the greenhouse effect, thanks to the efforts of a young Tennessee representative named Albert Gore Jr. As a college student, Gore had learned that humankind was on the brink of radically transforming the global atmosphere and risked bringing about the collapse of civilization. Gore, much like Pomerance, was stunned to learn this. If we really were facing catastrophe, why wasn't anyone talking about it?

As a representative, Gore decided that was going to change. Environmental and health stories had all the elements of narrative drama: villains, victims, and heroes. In a congressional hearing, you could summon all three, with the chairman serving as narrator, chorus, and moral authority. He wanted to hold a hearing every week. His staff, however, wasn't so sure this was a great idea. One colleague wondered whether the greenhouse effect would really attract the attention of the American people. "There are no villains," he said. "Besides, who's your victim?"

"If we don't do something," said Gore, "we're all going to be the victims."

What he didn't say was *If we don't do something, we'll be the villains too.*

AL GORE (1948–) served as a U.S. representative and senator before he became vice president under President Bill Clinton. His activism on the climate crisis began in the late 1970s and earned him the Nobel Peace Prize in 2007. "He is probably the single individual who has done most to create greater worldwide understanding of the measures that need to be adopted," the Nobel Committee said. His 2006 documentary on the  crisis, *An Inconvenient Truth*, won two Academy Awards. The companion audiobook won a Grammy. Gore wrote the 2017 documentary *An Inconvenient Sequel: Truth to Power* and continues to work in climate activism as founder of the Climate Reality Project. "Believe in the power of your own voice," Gore says. "The more noise you make, the more accountability you demand from your leaders, the more our world will change for the better."

James Hansen flew to Washington to testify at a congressional hearing organized by Gore on March 25, 1982. Gore had held another hearing the year before; as Gore's staff had feared, attendance had been very small: a few staffers and a bored reporter who left halfway through. Nevertheless he determined to try again, with Hansen serving as the star witness. Gore began by attacking the Reagan administration for cutting funding despite the "broad consensus in the scientific community that the greenhouse effect is a reality."

The other representatives at the hearing agreed. "How frequently must we confirm the evidence before taking remedial steps?" said one representative. "Now is the time," he said. "The research is clear."

Gore thought that they'd need even more answers before Congress would decide to restrict fossil fuel use. But others disagreed, urging immediate action. Hansen's job was to share the warning signs, to translate the data into plain English.

"We may already have in the pipeline a larger amount of climate change than people generally realize," said Hansen.

Gore asked when the planet would reach a point of no return—a "trigger point," after which temperatures would spike. "I want to know," he said, "whether I am going to face it or my kids are going to face it."

"Within ten or twenty years," said Hansen, "we will see climate changes which are clearly larger than the natural variability." In other words: Get moving on this, now.

A Nobel Prize—winning scientist on the witness panel added his voice: "It is already later than you think."

Now *this* hearing made some noise.

That night, the *CBS Evening News* spent nearly ten minutes on global warming. A correspondent explained that temperatures had increased over the previous century, that great sheets of pack ice in Antarctica were rapidly melting, that the seas were rising. "The trend is all in the direction of an impending catastrophe," the Nobel laureate said.

It seemed that something was starting to turn. The carbon dioxide issue was beginning to trouble the public consciousness— Hansen's own findings had become front page news, after all. What started as a scientific story was turning into a political story. Hansen and other scientists hadn't been so sure that was a good thing in the past—science was about facts, research, rigor. Politics was story and performance and spin. Now Hansen wondered whether this blending of the two might be a good thing. Maybe this was a way into a new future.

Then came news that the government report ordered by President Carter three years earlier was finally about to be published. A team of scientists had been asked to evaluate the consequences of global warming for the world order, and propose remedies. The Reagan White House had refused to take any action until it heard what these experts had to say. Finally, the big day arrived. The 496-page report, *Changing Climate*, was done. It was time to celebrate.

Rafe Pomerance wasn't invited to the fancy publication party. But he did snag an advance copy of the report. It started out strong: The introduction stated that action had to be taken immediately, before all the details could be known with certainty, or else it would be too late.

Yeah! Now they were getting somewhere.

Well . . . it said that, but . . . the authors' final conclusion was a little different. In fact they argued the opposite: There was no urgent need for action, and the public should not entertain the most "extreme negative speculations" about climate change.

Huh?

Yep. The message was *Slow your roll, people.* Better to wait and see. Better to bet on American ingenuity to save the day. Yes, the climate would change, mostly for the worse, but hey, future generations, maybe with their futuristic robot friends and hovercrafts, would be better equipped to change with it.

This message was echoed in newspapers across the nation: "A panel of top scientists has some advice for people worried about the much-publicized warming of the Earth's climate: You can cope."

The greatest blow, however, came from the *New York Times*, which published its most prominent piece on global warming to date under the headline "Haste on Global Warming Trend Is Opposed."

This was exactly the message that President Reagan's White House wanted the public to hear. *Let's not do anything just yet.* In the article, a science adviser to the president summed it up: "There are no actions recommended other than continued research."

Oh. Okay.

In the following weeks, press coverage withered, and the industry tuned out. The oil and gas companies decided they no longer had to worry about changing their business model—if the government wasn't going to act, why should they? Exxon, which had the industry's most sophisticated scientific operation, ended its carbon dioxide program. After all, if all those fancy official experts had concluded that emissions regulations were not a serious option, why should Exxon make a fuss?

Pomerance was frantic. He was not banking on technological advancement and good old American ingenuity. Technology had not solved the air and water crises of the 1970s. Activism and organization, forcing robust government regulation, had. The first lesson of activism was that you couldn't trust things just to get better by themselves. And things were bad: The climate problem was an *emergency*. A catastrophe! It was all right there, in great detail, on 496 pages of paper. To conclude that no action should be taken was not only insane, it was wrong.

Al Gore shared his desperation. Gore likened the greenhouse effect to "a bad science fiction novel," its ramifications so inconceivable—Manhattan as balmy as Palm Beach, Kansas posing as central Mexico—it seemed frivolous to keep talking about . . . talking some more.

At yet another hearing, Gore pulled together a slate of witnesses who he thought would attract the most media attention. This time he included Rafe Pomerance. It was unnatural for Pomerance to speak in public this way, but he was desperate. Word had to get out. He had no choice.

It is time to act. . . . We know what to do. The evidence is in. The problem is as serious as exists. People talk about not leaving this to their grandchildren. I'm concerned about leaving this to my children.

—**Rafe Pomerance,** Friends of the Earth, February 28, 1984

Pomerance had come with an action plan, which he entered into the congressional record: prepare for the climatic changes that were inevitable, fund more research, make conservation the highest consideration in all energy policy, and abolish the government program to create synthetic fuels from coal and oil. These measures would have the added benefit of reducing acid rain, increasing energy security, promoting public health, and saving money. "This issue is so big," said Pomerance, "yet the attitude that is being taken is so relaxed. I mean, it strikes one as a bit incredible."

He took a breath before concluding. "The major missing element in all this is leadership," he said. "It needs to come from the political community."

Would it?

Chapter 8
A HOLE IN THE SKY
CHANGES EVERYTHING

Breakthrough

Just when Rafe Pomerance had given up hope, he learned about a new study that seemed to have moved politicians to act. A team of British scientists had been studying ice conditions in Antarctica for years. After several years of results so alarming that they had disbelieved their own evidence, they reported a frightening discovery in the May 1985 issue of *Nature* magazine. The levels of ozone over Antarctica had declined drastically. Ozone shields Earth from ultraviolet radiation, acting as the Earth's "natural sunscreen."

Chlorofluorocarbons—chemical compounds containing carbon, chlorine, and fluorine—were responsible. CFCs were used in refrigerants and aerosol sprays because they were stable and a lot safer than the alternatives. So with refrigerators and air conditioners, cans of spray paint and superhold hairspray, we were blowing CFCs into the air at high rates. Problem was, the compounds were so stable that they lasted a long time and drifted high up into the atmosphere, where they finally broke down into chlorine and bromine—elements that destroy ozone.

WHAT'S SO BAD ABOUT THE OZONE HOLE?

Without our shield of ozone, NASA says, the sun's rays would sterilize the surface of the Earth. Yikes! "The layer is critical for the survival of all species, and without it life on land would not have evolved," says scientist Pawan Bhartia, an expert on the ozone layer. If our ozone shield weakens, more intense ultraviolet radiation would lead to quicker sunburns, cataracts in the eyes, skin cancer. The radiation would disturb plant and marine life, harming our food chain. Fortunately, since ozone-depleting chemicals were phased out around the world, the Earth's ozone layer is on its way to complete recovery.

The phrase used in the article was "a hole in the ozone layer." When most people heard this line, they pictured images like a pin stuck through a balloon, a crack in the ceiling. It was not exactly accurate—there was no "hole," but a seasonal depletion in levels of ozone. But the vivid language in headlines and on TV did succeed in worrying people. The "ozone hole" became a sudden global emergency.

Scientists understood that the ozone problem and the greenhouse gas problem were linked. CFCs were unusually potent greenhouse gases. Though the compounds had been mass-produced only since the 1930s, they were already responsible, by James Hansen's calculation, for nearly half of Earth's warming during the 1970s. But nobody was worried about CFCs because of their warming potential. They were worried about going blind.

When the alarm had first sounded in the 1970s, the United Nations began establishing a plan of action to address the problem. In 1985, they were ready to establish a framework for a global treaty to protect the ozone layer. Negotiators failed to agree on any specific CFC regulations, but two months later, after the British scientists reported their findings from Antarctica, President Reagan proposed a reduction in CFC emissions of 95 percent.

What a victory! It was a complete reversal.

Just a few months earlier, a large, powerful industry lobbying group had persuaded the Reagan administration that "there was too much uncertainty in the science to justify any further regulation of CFCs." But once the public discovered the ozone hole, every relevant government agency and U.S. senator urged the president to endorse the UN treaty. When Reagan finally submitted the Vienna Convention for the Protection of the Ozone Layer to the Senate for ratification, he praised the "leading role" played by the United States.

Yeah, okay.

Whatever the public relations spin was, it was a win for the environmentalists, including Rafe Pomerance. They'd take it. And it gave them an idea: What if a global treaty could do for the carbon dioxide problem what it had done for ozone? The United Nations had been holding semiannual conferences on global warming since the early 1970s, attended by scientists from all over the world. At the 1985 conference, as the world was growing alarmed about the ozone hole, an influential U.S. government scientist announced that he had changed his mind on climate change: "As a reversal of a position I held a year or so ago, I believe it is timely to start on the long, tedious, and sensitive task of framing a convention on greenhouse gases, climate change, and energy." The executive director of the UN Environment Programme, Mostafa K. Tolba, stood up and declared that it was time to begin serious consideration of "the costs and benefits of a radical shift away from fossil fuel consumption."

Though some warming was inevitable, the scientists acknowledged, the extent of the disaster could be "profoundly affected" by aggressive, coordinated government policies. Fortunately they had established, with the ozone treaty, a new international model to accomplish just that. The balloon could be patched, the ceiling replastered. There was still time. Maybe science and politics could work together to save the world.

Chapter 9
RUNNING OUT OF TIME

It's Coming from Inside the House

In September 1987, James Hansen saw signs of real progress and felt some momentum. Four years after *Changing Climate*, two years after an ozone hole had torn open the sky, and a month after the United States and more than three dozen other nations signed the Montreal Protocol to end the use of CFCs, the climate change crew was finally ready to party.

That year Hansen attended an event called "Preparing for Climate Change," which was meant to be a conference but felt more like a wedding. Scientists were there, politicians, and even representatives of the oil and gas industry. Everyone was hugging it out. Not even Hansen's presentation of his research could sour the mood. (The situation, remember, was pretty terrible.)

The political momentum had flipped. Now that the ozone problem was on the verge of being "fixed," climate issues had once again become a popular excuse for hearings on Capitol Hill—a noncontroversial subject that elicited concern, headlines, and lots of pompous speeches. In 1987 alone, there were eight days of climate hearings, in three committees, across both chambers of Congress; Senator Joe Biden from Delaware had introduced legislation to establish a formal national climate change strategy. Everyone thought that the issue would follow ozone's rise into signed and sealed international law. The head of President Reagan's EPA said as much.

It was because of all this good cheer that Hansen wasn't too concerned about a weird series of events that occurred just a week later. He was scheduled to testify at another Senate hearing, this time devoted entirely to climate change. On the Friday evening before his Monday appearance, he was informed that the White House demanded changes to his testimony.

Changes? Huh.

No rationale was provided. Nor could Hansen understand how the White House had the authority to censor scientific findings—the facts. But while the administration did have the authority to approve government witnesses, it couldn't censor a private citizen.

So at the hearing on November 9, Hansen listed himself as "atmospheric scientist, New York, NY." In his opening remarks, he said, "Before I begin, I would like to state that although I direct the NASA Goddard Institute for Space Studies, I am appearing here as a private citizen." He testified, and there were no other surprises.

Still, the brush with state censorship stayed with Hansen in the months that followed. He came to realize that people at the highest levels of the federal government—within the White House itself—hoped to prevent even honest dialogue about solutions. This was new—not just shrugs and indifference, but an active force against him, something like . . . a villain.

But on the surface, things looked pretty good. A New Deal for global warming—a comprehensive package of climate change legislation—was in the works. In June, President Reagan signed a joint statement with the Soviet Union that included a vague pledge to expand cooperation on global warming.

But a vague pledge didn't reduce emissions. There still wasn't any serious national or international plan to limit fossil fuel consumption. James Hansen was worried. No matter what anyone said, no matter what papers were signed or promises made, nothing had happened yet. And we were running out of time.

Then came the summer of 1988, and James Hansen wasn't the only one who could tell that time was running out.

The Heat Is On

The year 1988 brought the hottest and driest summer anybody had ever seen. Everywhere you looked, something was bursting into flames. Two million acres in Alaska were incinerated, and dozens of major fires scorched the West. Yellowstone National Park lost eight hundred thousand acres. The smoke drifted all the way to Chicago, twelve hundred miles away.

In Nebraska, suffering its worst drought since the Dust Bowl, there were days when every weather station in the state registered temperatures above 100 degrees. The director of the Kansas Department of Health and Environment warned that the drought might be the dawning of a climatic change that could turn the state into a desert in fifty years. In parts of Wisconsin, where the governor banned private fireworks and smoking cigarettes outdoors, the Fox and Wisconsin Rivers dropped so low that they were in danger of not being able to assimilate the waste discharged into them. "At that point," said an official from the state Department of Natural Resources, "we must just sit back and watch the fish die."

Harvard University, for the first time, closed because of heat.

New York City's streets melted, its mosquito population quadrupled, and its murder rate reached a record high.

Ducks fled the continental United States in search of wetlands, many ending up in Alaska, swelling its pintail population to 1.5 million from one hundred thousand.

It did not rain.

Texas farmers were forced to feed their cattle cacti (ouch).

Stretches of the Mississippi River flowed at less than one-fifth of normal capacity, causing massive, miles-long traffic jams of cargo ships. In the Cleveland suburb of Lakewood on June 22, yet another record smasher, a roofer working with 600-degree tar asked, "Will this madness ever end?"

That same day in Washington, DC, where it hit 101 degrees, Rafe Pomerance received a call from James Hansen, who was scheduled to testify before the Senate. Hansen explained that he had just received the most recent global temperature data. Barely over halfway into the year, 1988 was setting records. Already it had nearly clinched the hottest year in history. Ahead of schedule, the signal was emerging from the noise.

"I'm going to make a pretty strong statement," said Hansen. And the unbearably hot, dry conditions across the country might help make it even stronger. Hansen had told Pomerance that the biggest problem with the previous hearing (at least apart from the whole censorship thing) had been the month in which it was held: November. "This business of having global warming hearings in such cool weather," he said, "is never going to get attention." He wasn't joking. But they wouldn't have that concern this time.

Hansen worked on his testimony: "The global warming is now large enough that we can ascribe with a high degree of confidence a cause-and-effect relationship to the greenhouse effect." He wrote: "1988 so far is so much warmer than 1987, that barring a remarkable and improbable cooling, 1988 will be the warmest year on record." He wrote: "The greenhouse effect has been detected, and it is changing our climate now."

On the morning of the hearing, Hansen awoke to bright sunlight, high humidity, choking heat. It was signal weather in Washington, matching the hottest June 23 in history. It was 98 degrees outside when the session began at 2:10 p.m. The room was crowded with TV cameras and their hot lights, and senators who hadn't planned to attend showed up to make sure that they were seen. Hansen, the star speaker, was on.

IF 1988 WAS HOT, 2020 WAS . . . HOTTER

Depending on whom you ask, 2020 was either tied with 2016 for the warmest year on record, or just behind it. Either way, it's continuing a trend. The top ten hottest years have all occurred since 2005. Once a record breaker, 1988 doesn't even crack the top twenty-five today. Other grim facts recorded in 2020:

- Europe's and Asia's warmest years
- South America's second warmest year, North America's tenth
- Africa's and Australia's fourth warmest years
- Third-highest global ocean temperature, with parts of the Atlantic, Indian, and Pacific seeing record highs
- Smallest annual coverage of sea ice in the Arctic (tied with 2016)
- Most named tropical cyclones around the globe (103, tied with 2018)
- Fourth-smallest snow cover in the Northern Hemisphere

Hansen spoke calmly, wiping his brow, his eyes rarely rising from his notes. The warming trend, he said, could be detected "with 99 percent confidence." He encouraged the senators to do what was possible to curb fossil fuel emissions immediately.

But he saved his strongest words for after the hearing, when he was smothered by reporters. "It is time to stop waffling," he said, "and say that the evidence is pretty strong that the greenhouse effect is here."

Hansen's testimony led the national news; at the top of the front page, the *New York Times* headline read, "Global Warming Has Begun."

Woodstock for Climate Change vs. the Man

Finally, momentum!

Public awareness of the greenhouse effect reached an all-time high. Global warming caused one-third of Americans to worry "a great deal."

Presidential campaigns paid loud lip service to the cause. Candidate George H. W. Bush had strong words. "I am an environmentalist," he said on the shore of Lake Erie, the first stop on a five-state environmental tour. "Those who think we are powerless to do anything about the greenhouse effect are forgetting about the White House effect." His running mate emphasized the ticket's commitment to solving global warming at the vice presidential debate. "The drought highlighted the problem that we have and therefore we need to get on with it. And in a George Bush administration, you can bet that we will."

After Bush won the election, Washington insiders told reporters that his administration was going to act fast. They predicted that Congress would pass significant legislation once he took office.

By the end of the year, thirty-two climate bills had been introduced in Congress, led by two gigantic biggies: the Global Warming Prevention Act and the National Energy Policy Act of 1988. Both called for the establishment of an "International Global Agreement on the Atmosphere" by 1992. It seemed that a gasoline tax wouldn't be far behind. The United States was on the move.

Other countries were moving even faster. The West German parliament created a special commission on climate change; it eventually recommended a 30 percent reduction of carbon emissions. Sweden's parliament announced a national strategy to stabilize emissions and later imposed a carbon tax. The prime minister of England warned in a speech that global warming could "greatly exceed the capacity of our natural habitat to cope" and that "the health of the economy and the health of our environment are totally dependent upon each other."

The United Nations unanimously endorsed the establishment of the Intergovernmental Panel on Climate Change, to be composed of scientists and policymakers who would analyze the science and develop a global climate policy. World governments were teaming up, and the United States was chosen to lead the IPCC group working to find a response to climate change.

Chapter 10
PUSHBACK

Smoke and Mirrors

The oil and gas companies were worried. Rafe Pomerance's message was spreading; James Hansen's testimony before Congress had galvanized the American public, and politicians were making big promises about passing large, sweeping climate laws. The industry understood the science as well as anyone—it had been studying it for decades. So, how should it respond?

With a lot of questions.

Exxon's public relations strategists proposed that the company should emphasize the uncertainty of the scientific conclusions about the potential enhanced greenhouse effect. Yeah, okay, the experts were almost certain. *Almost.*

After strategizing with strategists and consulting with consultants, the American Petroleum Institute made a decision. The API would be "an active participant in the scientific and policy debate" and highlight uncertainties in the science, question the effectiveness of any new regulations, urge international cooperation, and accept only those measures "consistent with broader economic goals"—the ones that kept the money coming in. In other words, the trade group was going to undermine the message of Hansen, Pomerance, Gore, and other environmental leaders.

As the oil and gas industry prepared for the expected climate policy onslaught of the Bush administration, the API president tried out the new approach in a speech.

"Many people are already using the 'greenhouse' fever to push agendas built around extreme environmental and conservation ideas," he said. "Unless cooler, less biased heads prevail, the nation could scare itself into a costly, nearly impossible set of environmental goals, with tremendous burdens on U.S. industry and society."

He acknowledged the scientific consensus that carbon dioxide was rising and would warm the planet. But scientists couldn't say with certainty how quickly the warming would occur. It was important, he emphasized, that whatever the Bush administration did next, the industry had to stand together.

As it turned out, there was already someone deep inside the bowels of the White House who was working to undo all the promises that the president had made.

Inside Man

President Bush's chief of staff, John Sununu, had a piece of advice for members of the new administration: "Leave the science to the scientists. Stay clear of this greenhouse effect nonsense. You don't know what you're talking about."

So what was up with this Sununu guy? He had a reputation as a no-nonsense political wolf and had served three terms as New Hampshire governor. But Sununu still thought of himself as an "old engineer," as he was fond of putting it, having earned a doctorate in mechanical engineering from the Massachusetts Institute of Technology decades earlier. He couldn't quite be pigeonholed: He had fought angrily against local environmentalists to open a nuclear power plant, but as governor of New Hampshire, he had also signed an acid rain law, mandating a 25 percent reduction in sulfur dioxide

emissions, and lobbied President Reagan in person for a 50 percent reduction in sulfur dioxide pollution, the target sought by environmental activists. He increased spending on mental health care and public land preservation.

ACID RAIN is any form of precipitation, such as snow, rain, and hail, that is acidic enough to cause harm to the environment, such as to lakes and forests. In addition to carbon dioxide, the burning of coal and other fossil fuels releases sulfur dioxide and nitrous oxide. These gases mix with water in the atmosphere to form acids. Research suggests that global warming is likely to increase the formation of acidic precipitation.

In some ways Sununu was considered even more conservative than former President Reagan. He believed in conspiracy theories of socialistic forces using science to advance anti-democratic ideas. All the talk of long-term, far-reaching consequences of planetary warming sounded a little suspect to him.

So when Sununu learned that Al Gore, now a senator, was going to call a hearing to shame Bush into taking immediate action, he was ready for a fight. James Hansen would again serve as lead witness. Sununu reviewed Hansen's proposed testimony and was appalled: Hansen's warnings seemed extreme to him, especially since they were based on scientific arguments that Sununu considered, as he put it, "technical garbage."

At the same time, Bush's EPA chief, William Reilly, proposed that the president demand a global treaty to reduce carbon emissions. Sununu disagreed. It would be foolish, he said, to let the nation stumble into a binding agreement on questionable scientific merits, especially one that would compel some unknown quantity of economic pain.

And he made sure that his voice was heard.

Altered Statements

So, remember that time in 1988 when someone had tried to censor James Hansen's expert testimony? When he'd had to testify as a regular guy, and not the NASA employee that he was? It had bugged him a little, that episode, but he was getting ready to testify again, and it was about to get much worse.

In the first week of May 1989, Hansen received his proposed congressional testimony back from the White House. It was a shambles—lots of deletions and, more incredibly, some major additions.

Gore had called the hearing to increase pressure on Bush, but Hansen had agreed to testify for a different reason: He worried that one of his major points at the 1988 hearing had been misunderstood. His research had found that global warming would not only cause more heat waves and droughts like those of the previous summer; it would also lead to more extreme rain events. This was a crucial disclaimer—he didn't want the public to assume the next time there was a mild summer that global warming wasn't real.

But this proposed testimony, supposedly his own words, was now a mess. He told Senator Gore:

- The White House wanted him to basically deny his own scientific findings and call them "estimates" from models that were unreliable and "evolving."

- His anonymous censor wanted him to say that the causes of global warming were "scientifically unknown" and might be attributable to "natural processes," statements that would not only make his testimony meaningless but make him sound as if he had absolutely no idea what he was talking about.

- And in the most bizarre addition, he was asked to demand that Congress consider only climate legislation that would immediately benefit the economy, "independent of concerns about an increasing greenhouse effect"—a sentence no scientist would ever utter, unless maybe he was working for the American Petroleum Institute.

Remember the API? Emphasizing the sliver of scientific uncertainty? Supporting measures "consistent with broader economic goals"? The White House seemed to be doing the same thing.

James Hansen wasn't having it. When Senator Gore went to the press with this story of White House censorship, he had Hansen's full blessing.

The *New York Times* called Hansen the next morning. "I should be allowed to say what is my scientific position," Hansen said. "I can understand changing policy, but not science."

On Monday, May 8, the morning of the hearing, it was front page news: "Scientist Says Budget Office Altered His Testimony."

The official title of Gore's hearing was "Climate Surprises," but the only surprise the press cared about was the White House's interference with Hansen's testimony. In the overstuffed hearing room, the cameras were focused on the reluctant celebrity scientist.

Hansen held his statement in one hand and a single Christmas-tree bulb in the other—a low-budget prop to help explain that the warming already created by fossil fuel combustion was equivalent to placing a Christmas light over every square meter of the Earth's surface. After he read his "cleaned up" testimony, Gore pounced. He pretended to be surprised and puzzled by inconsistencies in Hansen's presentation. In a tone thick with mock confusion, he asked, "Why do you directly contradict yourself?"

Hansen explained that he had not written those contradictory statements. He did not quarrel with the White House's practice of reviewing policy statements by government employees. "My only objection is being forced to alter the science."

This was high drama, a made-for-TV moment.

Gore was beside himself. "The Bush administration is acting as if it is scared of the truth," he said. "If they forced you to change a scientific conclusion, it is a form of science fraud." He worked himself into a righteous fervor. "You know, in the Soviet Union they used to have a tradition of ordering scientists to change their studies to conform with the ideology then acceptable to the state. And scientists in the rest of the world found that laughable as well as tragic."

It was *on*.

Soon another scientist mentioned that the White House had tried to change his conclusions too. Gore called it "an outrage of the first order of magnitude."

The Senate hearing during the blazing hot summer of 1988 had created a hero out of James Hansen. Now, a year later, Gore had a real villain—a nameless censor in the White House.

After recess, the TV cameras and bright lights followed Hansen and Gore into the marbled hallway. Hansen insisted that he wanted to focus on the science. Gore was happy to focus on the politics. "I think they're scared of the truth," he said. "They're scared that Hansen and the other scientists are right and that some dramatic policy changes are going to be needed, and they don't want to face up to it."

At a press briefing later that morning, the White House admitted that Hansen's statement had been altered. The administration blamed an official "five levels down from the top" and promised that there would be no retaliation against Hansen, who was "an outstanding and distinguished scientist" and was "doing a great job." The episode did more to publicize the need for climate policy than any testimony Hansen could have delivered. It was "an outrageous assault" (*Los Angeles Times*) that marked the beginning of "a cold war on global warming" (*Chicago Tribune*), sending "the signal that Washington wants to go slow on addressing the greenhouse problem" (*New York Times*). The "White House effect" indeed.

Busted!

Science on the Sidelines

After this embarrassment, Sununu searched for ideas to rebuild the Bush administration's environmental rep.

We could begin, the EPA chief said, by recommitting to a global climate treaty. The United States was the only Western nation on record as opposing negotiations.

And so, three days after the disastrous Gore hearing, and two days after the British government had called on world leaders to organize a global warming convention as soon as possible, Sununu said America should work "to develop full international consensus on necessary steps to prepare for a formal treaty-negotiating process. The scope and importance of this issue are so great that it is essential for the U.S. to exercise leadership." He went on to propose an international workshop on global warming, to be hosted by the White House, that would aim to improve the accuracy of the science and calculate the economic costs of emissions reductions. Sununu was going all out.

It wasn't enough for Senator Gore, who had been trying to pressure President Bush to live up to his campaign promises: "Once again, the president has been dragged slowly and reluctantly toward the correct position," he said. "Although this is progress, it is still not nearly good enough."

Nor did it satisfy Rafe Pomerance, who told reporters that the belated effort to save face was a "waffle" that fell short of real action.

But the general public response was one of praise and relief.

Behind the scenes, Sununu wasn't happy. At all.

He'd decided to do his own research on the greenhouse effect, and he finished his one-man study more convinced than ever that Hansen's conclusions were nonsense. Hansen's models, Sununu groused, didn't begin to justify wild-eyed pronouncements that "the greenhouse effect is here" or that the 1988 heat waves could be attributed to global warming. No way should they be used as the basis for national economic policy.

Uh, okay.

His colleagues just nodded and smiled.

Sununu didn't even want to hear the issue discussed. "I don't want anyone in this administration without a scientific background using 'climate change' or 'global warming' ever again," he said on one occasion. "If you don't have a technical basis for policy, don't run around making decisions on the basis of newspaper headlines."

President Bush tried to keep a low profile and stay out of it. The truth was, the president had never taken a particularly strong interest in global warming. He had not discussed the issue at depth with scientists. (When scientists offered to brief the White House, they reported to Sununu.) Bush had only brought up global warming on the campaign trail because he had figured he needed a new issue to get him some positive press.

He trusted Sununu to figure it out for him.

Chapter 11
MISSED OPPORTUNITY

Hope

In spite of the setbacks engineered by Sununu and the American Petroleum Institute and others, an amazing chance arose to solve the global climate problem. In November 1989, four hundred representatives had been invited to the Dutch resort town of Noordwijk for a major diplomatic summit on global warming. To Rafe Pomerance, this was not just any old mundane meeting packed with empty speeches.

This was the moment.

If the world took this action, and lived up to the targets, we could actually solve climate change.

Every action taken by Pomerance, James Hansen, Al Gore,

and other activists had been building up to this. Their hope was that the conference would provide the formal ratification of Pomerance's most ambitious goal for global climate policy, rather than merely "a step closer to becoming global law." Representatives from sixty-seven nations would review the progress made by the Intergovernmental Panel on Climate Change and decide whether to endorse a framework for a global treaty. There was a general sense among the delegates that they would agree to the target proposed by the host, the Dutch minister—freezing greenhouse gas emissions at 1990 levels by 2000. At the very least, the delegates planned to approve a binding target of emissions reductions.

The environmentalists believed that the group *would* agree to a global freeze in the amount of greenhouse gas emissions and might agree to reduce emissions in future.

After more than a decade of fruitless international meetings, they would finally sign an agreement that meant something.

Woot! The energy among delegates was high.

Pomerance was filled with excitement, relief, anxiety. Victory seemed almost in hand. So what if he wasn't invited? It wouldn't have happened if not for his work over the last decade. He had to make it to the conference, even if that meant paying for his own travel, and holing up in a cheap hotel down the coast with friends from the Sierra Club, Friends of the Earth, and the Union of Concerned Scientists.

Skunks at the Garden Party

John Sununu was there too. He had been invited, as one of the official delegates for the United States.

Senators from both the Republican and Democratic parties were leaning on President Bush to act. Five Republican senators urged him to direct his negotiators to propose a "forceful and specific agenda" on global warming. "Unless you provide personal leadership on this issue," they wrote, "the United States will continue to send mixed signals to the world community and will put forth proposals that will be subject to criticism at home and abroad." A successful United States position had to include commitments to freeze U.S. carbon dioxide emissions at current levels, establish specific targets for reductions, and assist developing countries to use renewable sources of energy. If the government failed to enact "an aggressive domestic policy on carbon dioxide emissions," it could not expect other nations to act accordingly. The Republican senators called their proposal "the Bush plan" and offered the president permission to take credit for it. Forty Democratic senators sent their own letter the next week.

After that, Bush renewed his promise that the United States would "play a leadership role in global warming." Even Sununu seemed to have softened. In a speech to an international group of investors, he spent most of his time explaining why there had to be a coordinated international response to the threat of climate change. When an investor asked who would pick up the cost, Sununu replied: "Who picks up the cost if we don't fulfill our responsibilities as stewards of the environment? . . . The long-term cost will be less by doing it right now than it will be by trying to retreat from a disaster fifty or a hundred years from now."

The old engineer seemed to be singing a new song!

Pomerance wondered whether it was enough. He had recently commissioned a report tracking global greenhouse gas emissions, and the findings were troubling. The United States was the largest contributor by far, producing nearly a quarter of the world's carbon emissions, and its contribution was growing faster than that of every other country. Bush's indecision, or inattention, had already managed to delay ratification of a treaty until 1990 at the soonest, perhaps even 1991. By then, Pomerance worried, it would be too late.

Incredible as it seemed to Rafe Pomerance, it had been a decade since he helped warn the White House of the dangers posed by fossil fuel combustion.

It was nine years since his first desperate efforts to write legislation, reshape American energy policy, and demand that the United States lead an international campaign to arrest climate change.

It had been a year since he devised the first emissions target proposed at a major international conference.

And now, at the dawning of the 1990s, senior officials from more than sixty nations were debating the merits of a binding global treaty. But Pomerance was on the outside, not in the halls of power. As he stared at the wall separating him from the ministers' muffled debate, he could only hope that all his work had been enough.

Pomerance and his friends planned to stage a stunt each day to galvanize support for a binding treaty. The first took place at the flagpoles, where they met a photographer from a European news service. Performing for the photographer at dawn, they lowered the Japanese, Soviet, and American flags to half-staff. Then they gave a reporter an outraged statement, accusing the three nations of conspiring to block the one action necessary to save the planet. The article appeared on front pages across Europe.

The main meeting began in the morning and continued into the night, much longer than expected; most of the delegates had come to Noordwijk prepared to sign the Dutch proposal. To use the bathroom, the diplomats had to exit the conference room and negotiate the hallway, squeezing past the activists; each time the doors opened and a minister darted out, Pomerance and the other activists leaped up, demanding an update.

As the night went on, and on, the word got out: The United States was ruining everything.

When, close to dawn, the beaten delegates finally emerged, Pomerance learned what had happened. Sununu had won. With the acquiescence of Britain, Japan, and the Soviet Union, the U.S. representatives had forced the conference to abandon the commitment to freeze emissions. The final statement noted only that "many" nations supported stabilizing emissions—but it did not indicate which nations, or at what level, or by what deadline.

And with that, a decade of excruciating, painful, exhilarating progress turned to air.

Wait.

WHAT?!

Bitterly disappointed activists called the United States and other dissenting nations the "skunks at the garden party," stinking up the summit with their science-denying, money-grubbing ways. In Washington, Al Gore mocked Bush on the floor of the Senate: For all the brave talk about "the White House effect," the president was practicing the "whitewash effect."

Pomerance tried to be more diplomatic. "The president made a commitment to the American people to deal with global warming," he told the *Washington Post*, "and he hasn't followed it up." He didn't want to sound defeated. "There are some good building blocks here," he said, and he meant it. The Montreal Protocol, the ozone agreement, wasn't perfect at first either—it had huge loopholes and weak restrictions. Once in place, however, the restrictions could be tightened. Perhaps the same could happen with climate change.

Perhaps.

Possibly.

Maybe.

Pomerance wanted to believe that this was progress. To do so, however, he'd have to forget everything he'd learned since opening the pages of the coal report in 1979. He had been brave enough to tell the truth about the Earth's future to Congress, to three presidents, to the world. But there was a limit to what he dared to tell himself.

Part III

HANDLING THE TRUTH

For a long time we were able to hold ourselves
in a distance, listening to data and not being
affected emotionally . . . But it's not just a science
abstraction anymore. I'm increasingly seeing
people who are in despair, and even panic.

—Lise Van Susteren, *cofounder, Climate Psychiatry Alliance*

We may encounter many defeats, but we must not
be defeated. It may even be necessary to encounter
the defeat, so that we can know who we are.

—Maya Angelou, *writer, performer, and activist*

Chapter 12
THAT WAS THEN, THIS IS NOW
(AND THEN)

Blame Game

So here we are. The United States had a chance to do something great and, by all accounts, blew it. The sad truth is this: Since the failed November 1989 conference about the dangers of global warming at the Dutch resort town of Noordwijk, more carbon has been released into the atmosphere than in the entire history of civilization preceding it.

Why? Whose fault was it?

It was true that the money guys had been talking. They wanted economists and businesspeople, not environmentalists, to dictate the terms of any treaty to reduce emissions of CO_2. Bush's entire economic council had come together behind the position that the benefits of emissions cuts should be weighed against immediate economic costs. And those costs needed to be calculated—slowly and deliberately.

And of course, there was the "old engineer," John Sununu. In 1989, was George H. W. Bush's chief of staff the only person standing in the way of a binding international agreement to prevent catastrophic global warming?

"Yes," says William Reilly, Bush's EPA chief. "And no."

He points out that Sununu's obstruction was critical at a time when the momentum was good. Public support for climate policy was at an all-time high, the international process received vocal bipartisan support, and a binding treaty along the terms broadly agreed on at Noordwijk would have kept planetary warming to 1.5 degrees.

But he's not completely to blame. The negotiation of the first international accord on climate change continued for another two and a half years before it was finalized at the 1992 Earth Summit in Rio de Janeiro. The largest gathering of world leaders in history at the time managed to produce only a nonbinding agreement to reduce greenhouse emissions, with no mandatory timetables or targets. At any point Bush could have demanded a binding treaty, and likely forced one: With the dissolution of the Soviet Union at the end of 1991, the United States not only dominated the world order economically and militarily, but it was responsible for about a quarter of humanity's carbon emissions every year. Although his chief scientist was urging him to consider strong restrictions and carbon legislation, Bush did not. And in the final, critical six months of negotiations of the climate accord, John Sununu was not there: He resigned as Bush's chief of staff in December 1991, under a cloud of criticism for his abrasive style, not to mention his use of military jets to take him skiing or to the dentist.

But even if Sununu can take the sole blame for the collapse at the Noordwijk summit, that brings up another, unsettling question: Why was support for a climate remedy so shallow that all it took was a single naysayer within the Bush administration to unravel it?

Today, John Sununu himself believes that the collapse would have happened whether or not he had objected. "The leaders in the world at that time were all looking to seem like they were supporting the policy without having to make hard commitments that would cost their nations serious resources. That was the dirty little secret at the time."

To Sununu, the international process was all a sham, designed to make countries look like they cared. Sununu is convinced that even if the United States had signed a strict treaty, it wouldn't have had any bearing on emissions levels. "The other nations were saying we'll ride the horse, and since we don't have to make any commitments, we can look like we're on board."

Maybe Sununu is right. When delegates from the major European nations were asked how they intended to stabilize their greenhouse gas emissions, they could not answer. "Who knows?" one of them said. "After all, it's only a piece of paper and they don't put you in jail if you don't actually do it."

That's the problem with "binding global treaties." It's not like there's a global police force about to arrest entire countries for breaking their agreements. No one really wants to get bogged down in military action or economic penalties for failing to meet emissions targets. We all have to monitor ourselves, give ourselves consequences. And if we're willing to do that, what do we need a binding treaty for?

Here's the thing: Our understanding of the global warming problem hasn't changed substantially during this time. The science only gets more precise.

The political story hasn't changed much either, except in its particulars. Sununu is correct that to this day even some of the nations that have advocated most aggressively for climate policy— among them the Netherlands, Canada, Denmark, and Australia— have failed to honor their own commitments. Only seven countries are close to limiting emissions at the level necessary to keep warming to two degrees: India, the Philippines, Gambia, Morocco, Ethiopia, Costa Rica, and Bhutan.

When it comes to the United States, which has not agreed to make any binding commitments whatsoever, the main story for the last quarter century has been the full-throttle efforts of the fossil fuel industry, with the help of the Republican Party, to suppress scientific fact, confuse the public, and bribe politicians.

Turning Lies into Truth

Remember the American Petroleum Institute? In July 1989, the API drafted its "Position on Global Climate Change." It warned against "premature" policies "based on today's limited understanding of the issue," while advocating for "measures that will reduce the threat of climate change, yet also make sense in their own right."

Say what now?

This gobbledygook translates into a two-pronged mission statement for the oil and gas industry and its allies: Let's pretend to do what we say, and keep doing what we do.

In other words, sow doubt about the science, and if policy changes are coming, let's try to keep them as minor and ineffective as possible so that they don't mess with the money.

After the famous Hansen hearing in 1988, the API established a lobbying organization, ironically called the Global Climate Coalition. The U.S. Chamber of Commerce joined in, and so did trade associations representing the coal, electric power, and automobile industries. At first, the coalition shared and reacted to news of any proposed regulations, but then it began to create its own news in a very effective press campaign. Lobbyists gave a few briefings to politician buddies and contacted scientists who had expressed doubts about global warming, paying them to write opinion pieces. Soon their statements were in the national press.

"You know very well we can't predict the weather with any certainty," said one skeptical scientist with a chuckle in a major wire-service story that reported "many respected scientists say the available evidence doesn't warrant the doomsday warnings."

The *New York Times* published a letter from another scientist claiming that there was "considerable doubt in the scientific community about the greenhouse climate warming."

Forbes, a national business magazine, devoted its cover to "The Global Warming Panic," and even *Newsweek*, which had published frequent reports on the subject for two years, was prompted to ask, "Is It All Just Hot Air?"

Even though an outpouring of articles proclaimed "doubt in the scientific community," the highly respected journal *Science* in 1991 placed the total number of "outspoken greenhouse dissidents" in the United States at "a half-dozen or so." But the appearance of widespread disagreement was enough.

President Bill Clinton, with Al Gore as vice president, came into office in 1993. He proposed an energy tax, in the hope of meeting the goals of the 1992 Rio treaty. The Global Climate Coalition did not approve, and the proposal was defeated by a well-funded disinformation campaign. Senate Democrats from oil and coal states joined Republicans to defeat Clinton's tax proposal, and through the rest of the decade, the coalition spent at least $1 million a year to crush public support for climate policy.

The international effort to come up with binding targets to reduce greenhouse emissions continued, however, ahead of a 1997 summit in Kyoto, Japan. Rafe Pomerance was even officially invited to that one. By then he had been named deputy assistant secretary in the State Department's Bureau of Oceans and International Environmental and Scientific Affairs. But every effort by the U.S. delegation to win support for emissions reductions and carbon trading brought poisonous attacks from industry and the Republican Party, coordinated by the Global Climate Coalition.

President Clinton and Vice President Gore, though they supported the delegation's efforts, failed to win over the opposition even within their own administration, particularly among their economic advisers.

The U.S. delegation endorsed the Kyoto Protocol, which committed countries to reducing greenhouse gas emissions over a five-year period by an average of 5 percent below 1990 levels. But the treaty was never submitted to Congress for ratification. After the Global Climate Coalition spent $13 million on a single ad campaign, the Senate voted on a preemptive resolution declaring its opposition to a treaty. The resolution passed 95–0.

There has not been another serious effort to negotiate a binding global climate agreement since.

With its remarkable record of success, the API's Global Climate Coalition made a bold shift in its message. Rather than emphasizing uncertainty in the "magnitude and timing" of climate change, the coalition went even further, saying that *the fundamental science of climate change*—the science that was established in the nineteenth century, that had been validated by National Academy of Sciences in 1979 and confirmed by every major study since—*was itself uncertain*. It was deceptive double-talk, like a historian arguing that slavery was not the primary cause of the Civil War . . . then arguing that slavery did not exist.

Like all cartoonishly evil bad guys, the Global Climate Coalition finally fell apart in 2002. They were even embarrassed by themselves; one senior oil industry employee said, "We didn't want to fall into the same trap as the tobacco companies who have become trapped in all their lies." Besides, the truth was that their work had been done so well that they didn't need to do it anymore. In the 2000 presidential election, George W. Bush and Dick Cheney, a former CEO of the oil giant Halliburton, had beaten Gore and won the White House.

The head-in-the-sand approach continued. Although President George W. Bush acknowledged that climate change was real and vowed to reduce greenhouse gas emissions, nothing happened. In 2008 Republican presidential nominee John McCain called for a mandatory limit on U.S. emissions, but much of the Republican Party took a position of denial after Democrat Barack Obama, the first Black president of the United States, took office in 2008. In Congress, the House managed to pass comprehensive climate legislation in 2009, but the Senate, even with a Democratic majority, declined to take it up. In that year alone, the oil and gas industry spent about half a billion dollars on lobbying efforts to weaken energy legislation. The largest donor to that lobbying campaign was Exxon-Mobil.

Who Knew? We All Did.

Remember Exxon? The Exxon that was at the forefront of the climate science charge? All the years of big talk and well-financed questionable action have made Exxon an especially vulnerable target for lawsuits seeking compensation for the damage that's been done.

In recent years, lawsuits claiming harm from companies' actions, or negligence, have emerged as a possible way to force movement on the climate crisis, as the science of assigning regional effects to global emissions has grown more precise. Other cases have sought to force climate action through laws already on the books such as the Clean Air Act, the National Environmental Policy Act, and the Endangered Species Act, and through long-established legal doctrines such as the principle that government must preserve and protect certain natural resources for public use. This is one part of climate science that has advanced significantly since 1979—deciding whom to blame.

The rallying cry of the legal battle is "Exxon knew." And that's true: Senior employees at Exxon, and its predecessor, Humble Oil, like those at many other major oil and gas corporations, knew about the dangers of climate change at least as early as the 1950s and did nothing to reduce emissions.

But the automobile industry, responsible for nearly 20 percent of U.S. carbon emissions, knew too.

The electric utilities, whose coal- and gas-burning power plants are responsible for 31 percent of CO_2 emissions in this country, have known since the 1970s, when their trade research association began conducting studies on the subject. They all own responsibility for our current paralysis and have made it much more painful than necessary. But they haven't done it alone.

The U.S. government knew. Every president since John F. Kennedy in the early 1960s has considered the merits of acting on climate policy. Congress has been holding hearings for forty years; the intelligence community has been tracking the crisis even longer.

The preeminent scientific journals *Nature* and *Science* have been publishing climate change studies for nearly a half century.

The environmentalists knew too—items appeared in the newsletters for the Sierra Club and the Natural Resources Defense Council in the late 1970s. With the exceptions of Rafe Pomerance and Friends of the Earth, and the World Resources Institute, however, there was no widely known sustained effort by activists to address the crisis until the late 1980s.

A LOT of people knew.

Way back in the 1950s, *Time* magazine, the *New York Times*, and *Popular Mechanics* all ran articles on the increased carbon dioxide in the atmosphere due to the burning of coal and oil. A Johns Hopkins physicist predicted the Earth would warm by as much as 1.5 degrees Celsius in a century. Scientist Roger Revelle said all that CO_2 could have violent effects on the climate. The *New York Times* science editor at the time saw where things were headed: "Coal and oil are still plentiful and cheap in many parts of the world," he wrote, "and there is every reason to believe that both will be consumed by industry so long as it pays to do so." *Time* wondered whether "man's factory chimneys and auto exhausts will eventually cause salt water to flow in the streets of New York and London."

THIS IS BAD

In 1958, the Bell Science Series of television specials aired *The Unchained Goddess*, a film about weather, produced by Frank Capra, best known for the holiday classic *It's a Wonderful Life*.

Toward the end of the film, the kindly Dr. Research says, "Man may be unwittingly changing the world's climate" because of all the carbon dioxide released from factories and automobiles.

"This is bad?" asks his costar.

"A few degrees' rise in the Earth's temperature would melt the polar ice caps," Dr. Research narrates off-screen, as glaciers collapse like downed skyscrapers and an animated sightseeing boat floats over underwater ruins. "And if this happens, an inland sea would fill a good portion of the Mississippi Valley. Tourists in glass-bottomed boats would be viewing the drowned towers of Miami through one hundred and fifty feet of tropical water. For, in weather, we're not only dealing with forces of a far greater variety than even the atomic physicist encounters, but with life itself."

Yeah, this is bad.

Everyone knew—and we all still know. We know that the trans-formations of our planet, which will come gradually and suddenly, will reconfigure the political world order. We know that if we don't sharply reduce emissions, we risk the collapse of civilization.

We also know that the coming changes will be worse for today's young people, worse yet for the next generation, and so on. So far our actions say that those lives mean nothing to us.

It can't be that bad, right? We can't be that bad! The situation must not be quite so terrible as that, we might say, shaking our heads in disbelief. Surely there's time for a sensible transition to renewable energy—we'll get to it soon. Of course we care about our grandchildren!

We do not like to think about loss or death. Americans, in par-ticular, don't like to think about death. We don't want to face the worst. Our weak, vague language—*global warming, climate change*—makes that obvious.

If You Don't Know, Now You Know

Now we know the political story of the climate crisis, the tech-nological story, the economic story, the industry story. They are all critical to understanding how we got here. But what about the human story? How do we live with the knowledge that the future will be far less hospitable than the present? What do our failures say about our substance as a people, as a society, as a democracy?

Listen to us now: "I believe man has an impact on the climate, but what is not completely understood is what the impact is." "We don't understand what the effects [of climate change] are." "I would not agree that [carbon dioxide]'s a primary contributor to the global warming that we see." These misleading statements come from Cabinet members of the administration of President Trump in 2017.

A NONBELIEVER IN THE WHITE HOUSE

Republican President Donald Trump was the most vocal advocate against climate change science and legislation who ever occupied the White House. As president from 2017 to 2021, he succeeded in scaling back or eliminating over a hundred environmental rules and regulations: weakening limits on emissions from cars and trucks, eliminating a requirement for oil and gas companies to report on methane emissions, and relaxing air pollution regulations for power plants that burn waste coal.

In countless tweets, speeches, and interviews he expressed his skepticism about global warming, which he had made clear in his 2016 presidential campaign:

"I think there's a change in weather. I am not a great believer in man-made climate change. I'm not a great believer. There is certainly a change in weather that goes—if you look, they had global cooling in the 1920s and now they have global warming, although now they don't know if they have global warming. They call it all sorts of different things; now they're using 'extreme weather' I guess more than any other phrase . . . I am not a believer. Perhaps there's a minor effect, but I'm not a big believer in man-made climate change."

The words of these politicians are exactly as honest as claims that cutting taxes on the rich will help the poor or that cigarette smoking helps digestion. They are lies. But hey, if the science is "uncertain," no one can be blamed for doing nothing. Our collective failure to respond to the crises heightened by rising temperatures, Pope Francis says, "points to the loss of that sense of responsibility for our fellow men and women upon which all civil society is founded."

It seems that we've decided to just . . . worry about it later, figure it out as we go along. We have favored adaptation over mitigation.

ADAPTATION means adjusting to changing circumstances, reacting to what's already happening. With climate change, it means figuring out how to live with the effects of climate change and even use it to our advantage (like longer growing seasons). *Mitigation* means reducing the harmful effects of something, trying to prevent the worst from happening in the first place. To mitigate climate change, we have to reduce the levels of greenhouse gases in the atmosphere and slow the rate of global warming.

There are signs of hope.

In 2020, Trump lost the presidential election to his challenger Joe Biden, who has pledged an aggressive rollback of Trump's destructive environmental policies. In July 2021, as temperatures in the Pacific Northwest hovered at 40 degrees warmer than average, Biden sarcastically addressed the falsehoods pedaled by climate science skeptics. "Anybody ever turn on the news and see it's a hundred sixteen degrees in Portland, Oregon? A hundred sixteen degrees. But don't worry, there's no global warming. Doesn't exist. It's a figment of our imagination. Seriously."

In November 2021, after months of hard-fought negotiations, President Biden signed a $1 trillion infrastructure bill into law, including $47 billion for what was called the United States' "first major investment in climate resilience." The Build Back Better bill includes financial support for communities vulnerable to climate-related disasters, and funding to increase access to clean drinking water, invest in fossil fuel alternatives, and support electric vehicles. But many, including Rhode Island Senator Sheldon Whitehouse, warned that there was more to be done. "It's significant that we could get a significant bipartisan measure that recognized that climate change was real and we need to protect our infrastructure against its impacts," said Senator Whitehouse. "But it's not enough to just do repair work. We need to prevent the worse scenarios."

Since 1979, when environmentalist Rafe Pomerance first read about global warming, many solutions have been developed to turn things around, such as carbon taxes, renewable energy invest-ment, replacing gasoline-burning vehicles with electric vehicles, expansion of nuclear energy, reforestation, improved agricultural techniques, and even machines capable of sucking carbon out of the atmosphere. But what is missing is a global commitment to the solution. Humankind is capable of such a commitment—the worldwide vaccination effort to end the COVID-19 pandemic has shown that we can take on a monumental fight.

Pomerance's favorite partner in the fight to alert the govern-ment and the public about the threat of global warming, scientist

James Hansen, hasn't given up. "We're still at a point where we could deal with this problem, but only if we begin to make the price of fossil fuels honest, so that we begin to move to clean energies, but we're not doing that. We just have governments talking about, 'Oh, they'll have some goal for the future,' but that doesn't do anything. As long as fossil fuels seem to be the cheapest energy to the public, they will keep using them. [And we] have to encourage the technologies that are needed for clean energy. And frankly, to go to carbon-free electricity, I think most nations will probably need to use next-generation safe nuclear power."

Rafe Pomerance hasn't given up either. He has devised his own practical solution to climate change—not a technological solution but a political one. He argues that the critical legislative body for curtailing global emissions is the U.S. Congress. If it insists on major climate policy, he believes, the rest of the world will follow. How, then, to motivate congressional action? It is the problem he has been working on, more or less, since 1979. Pomerance is now a consultant for ReThink Energy Florida, which hopes to alert the state to the threat of rising seas. Republican congressmen in Florida have a healthier fear of climate change than their colleagues—a reasonable position in the state that, by a wide margin, is most imperiled by sea level rise. Pomerance believes that if he can persuade Florida Republicans to demand policy action, they can help turn the rest of their party.

When he feels despondent, he wears a bracelet that his granddaughter made for him, to remind him why he continues to fight.

Chapter 13
TIME FOR CHANGE

Finding New Hope in Old Ways

In previous chapters, we've seen that many governments, institutions, and politicians (mostly white men) have taken a head-in-the-sand approach to the climate crisis and environmental justice. But again, there have always been those who believe, as anthropologist Margaret Mead did, that "a small group of thoughtful, committed citizens can change the world." And many of those citizens today are young people committed to equity and inclusion in their work, and who reflect the racial, ethnic, gender, and geographic diversity of our world. They work toward the future by looking to the lives and work of their ancestors for solutions.

> We are at a tipping point right now where we will either be remembered as a generation that destroyed the planet, as a generation that put profits before future, or as a generation that united to address the greatest issue of our time by changing our relationship with the earth. We are being called upon to use our courage, our innovation, our creativity, and our passion to bring forth a new world.
>
> —**Xiuhtezcatl Roske-Martinez**, Indigenous climate activist

Youth-led groups have raised public awareness of the cascading damage of climate change through social media, mass demonstrations, petitions, and contact with government officials. "Our ancestors are the ones that died fighting for this land, so that makes me think that we have a duty to fight for our land. And we are obligated to protect the soil and the water and everything that is sacred like that," said Tokata Iron Eyes, a young Lakota environmentalist. "Whatever happens with . . . climate change—that is going to be affecting us, this generation. And it will affect the next generation too."

In 1990, Indigenous activists and youth from various tribal Nations met to discuss issues of environmental justice, and the Indigenous Environmental Network was founded. With continued gatherings over the years to focus on the protection and nurturing of "Mother Earth," the network works to recognize and implement practices based on traditional knowledge and respect for nature, to involve people of all ages in environmental justice work, build sustainable and healthy Indigenous communities, and get involved in policy decisions that affect Indigenous land and communities.

Toxic waste dumping, use of pesticides, deforestation, mining, water rights, and the impact of the climate crisis are among the issues that the group works on. IEN's advocacy includes facilitating dialogue between tribal governments, members, and young people, organizing public information campaigns and protests, and providing financial assistance and community-based advisers to support local environmental justice work. "The Indigenous Peoples of the Americas have lived for over 500 years in confrontation with an immigrant society that holds an opposing world view," declares the network's Unifying Statement of Principles. "As a result we are now facing an environmental crisis which threatens the survival of all natural life."

Indigenous activism for climate and environmental justice was at the forefront of the Dakota Access Pipeline protests in 2016. The pipeline runs 1,172 miles from oil fields in North Dakota to a terminal in Illinois, just north of the Standing Rock Sioux Reservation, where members of the Lakota and Dakota nations live.

It can transport 570,000 barrels of crude oil per day. In legal challenges and public demonstrations, members of the tribe and their supporters have argued that they were not adequately consulted about the route. "This is an environmental justice issue, where the impacts on and input from a more affluent, white community had priority over the impacts on and input from Indigenous Peoples," declared the Grassroots Global Justice Alliance.

Indigenous peoples are living with the overlapping effects of colonialism, industrial exploitation and climate change. However, we retain a powerful commitment to protecting all that generates life, both within our ancestral practices and the life-giving forces within the land. This is why Indigenous peoples throughout the world—and primarily Indigenous women and girls—are at the very forefront of movements for decolonization and climate justice.

—**Siku Allooloo,** Indigenous writer and community organizer

The pipeline travels under the reservoir that provides Standing Rock's drinking water. The community raised concerns that the "Black Snake" could jeopardize their water supply and that construction would damage sacred sites, violating their tribal treaty. The pipeline's delivery of a half million barrels a day would also contribute fifty million tons of CO_2 each year to the atmosphere. Opposition to the project brought thousands of protesters to Standing Rock, from hundreds of Indigenous nations—from every state in the United States and from countries as far away as Tibet, Sweden, Guatemala, and Brazil.

Women were at the forefront of the opposition movement. LaDonna Brave Bull Allard, tribal historian of the Standing Rock Sioux, established one of the more well-known Standing Rock campsites, Sacred Stone. "Women were the people who held the tribe together and they were the willpower of the tribe and its strength. So, just knowing that we come from such powerful genes makes us feel strong inside," said Tokata Iron Eyes, twelve years old at the time. "We as young women have an obligation to fight for ourselves and our people. I feel like I need to fight for my kids and my grandchildren, and my grandmas who can't fight for themselves anymore."

Months of public demonstrations, calls for boycotts, and legal challenges drew attention and support from celebrities, environmental organizations, and activists around the world—as well as rubber bullets, tear gas, and fire hoses from police. President Obama finally shut the project down. The "water protectors"— who see themselves performing a sacred duty to protect and preserve the Earth's water—had won, but only temporarily.

In 2017, President Donald Trump issued an executive order for the pipeline to continue. Protesters were ordered to evacuate, and some were arrested. The pipeline began operating in June 2017, despite a continuing battle in the courts. In July of 2020, a U.S. judge ordered a shutdown of the pipeline, calling for an extensive review of its environmental impact, but his order was overturned and the pipeline continued to operate, without a federal permit. In

April of 2021, amid protest from Indigenous leadership and activists, the Biden administration announced that it would not stop operations during the review.

Group Project

Top-down approaches to climate problem-solving are being rejected around the world. The Pan African Climate Justice Alliance is a group of over a thousand organizations from forty-eight African countries working together in the fight against the climate crisis. By bringing together grassroots organizations, community members, and what it calls the voices of those at the frontline of the climate crisis, the alliance takes a partnership approach that respects the culture and knowledge of the people it looks to serve.

On World Environment Day in 2020, the alliance issued a call to action: "We must not sit pretty as the seas and air are so polluted that breathing almost becomes a crime . . . a death sentence, and desertification, deforestation, and loss of biodiversity expose us to hunger, famine, unreliable and unpredictable rain patterns and in a lot of cases land/mudslides and lately even cyclones. Those that turn our lives upside down in the twinkling of an eye."

A lack of political will among national governments has led to climate policy gridlock. But many local communities are focusing on solving their own climate problems, and we can learn from them.
—**Ron Brunner,** scientist, University of Colorado

Cities That Never Sleep

While big and powerful governments flounder at the national level, cities and states are taking matters into their own hands. In 2019, New York City declared a "climate emergency," joining cities like its next-door neighbor, Hoboken, New Jersey, as well as Darebin, Australia, and London, England. In California, Los Angeles even established a Climate Emergency Mobilization Office. These efforts are not new.

"Beginning in 1998, the Danish island of Samsø achieved carbon neutrality [a carbon footprint of zero] in about five years, partly through wind turbines funded by the islanders themselves," reports NASA. "Samsø now generates more electricity from renewable resources than the island needs and sells the surplus energy to the mainland, replacing carbon dioxide emissions from fossil fuel consumption while growing its economy."

The Green New Deal

One of the most significant developments in climate activism in the last three decades is the Green New Deal. The proposal is designed to be the most ambitious national project taken on since President Franklin D. Roosevelt's New Deal during the Great Depression. As Roosevelt's programs created a broad mobilization effort to pull the United States out of the economic crisis in the 1930s, so the Green New Deal is meant to mobilize society and our economy to tackle the climate crisis today. Like the original New Deal, the proposal calls for government investments in communities, public works projects, and private industry.

The idea of a broad mobilization to reduce carbon emissions has been around for years, but it wasn't widely known until 250 young activists with the Sunrise Movement staged a sit-in at Representative Nancy Pelosi's office in 2018. Democrats had flipped the House in the November midterm elections, and Pelosi was set to take over as Speaker. The timing was right to demand that Congress take broad action on the climate crisis in a Green New Deal.

On February 7, 2019, newly elected Representative Alexandria Ocasio-Cortez of New York and Senator Edward Markey of Massachusetts submitted resolutions to the House and Senate "recognizing the duty of the Federal Government to create a Green New Deal." Sixty-five representatives and eleven senators cosponsored the resolution when it was originally introduced. It now has over a hundred cosponsors in the House and Senate—all Democrats plus one independent, no Republicans. The goal is to move the United States completely away from the use of fossil fuels and, in the process, transform the U.S. economy and way of life.

Because the United States has historically been responsible for a disproportionate amount of greenhouse gas emissions, having emitted 20 percent of global greenhouse gas emissions through 2014, and has a high technological capacity, the United States must take a leading role in reducing emissions through economic transformation.
—HR 109 and SR 59, Green New Deal resolutions introduced in 116th Congress

For such a bold, complex proposition, the language of the resolution is surprisingly simple and straightforward. The resolution states that human activity is the dominant cause of observed climate change over the past century. It also states that a changing climate is causing sea levels to rise and an increase in wildfires, severe storms, droughts, and other extreme weather events that threaten human life, healthy communities, and critical infrastructure.

The five main goals of the Green New Deal are clearly summarized:

- Achieve net-zero greenhouse gas emissions through a fair and just transition for all communities and workers
- Create millions of good, high-wage jobs and ensure prosperity and economic security for all people of the United States
- Invest in the infrastructure and industry of the United States to sustainably meet the challenges of the twenty-first century
- Secure clean air and water, climate and community resilience, healthy food, access to nature, and a sustainable environment for all
- Promote justice and equity by stopping current, preventing future, and repairing the historic oppression of frontline and vulnerable communities

The resolution's definition of "frontline and vulnerable communities" is long, reflecting the diversity of people who have been harmed or ignored by our government throughout history: "indigenous peoples, communities of color, migrant communities, deindustrialized communities, depopulated rural communities, the poor, low-income workers, women, the elderly, the unhoused, people with disabilities, and youth."

As the youth-led Sunrise Movement has said, "The Green New

Deal is the only plan put forward to address the interwoven crises of climate catastrophe, economic inequality, and racism at the scale that science and justice demand."

A Black research institute in Chicago, the New Consensus, helped develop the Green New Deal. The think tank's policy director, Rhiana Gunn-Wright, was instrumental.

"Growing up I'd wonder about structures—in my neighborhood and schools. What are the rules, who made the rules? You can't just look at the surface," Gunn-Wright says. She works on government policy so she can create solutions to complex problems that will directly benefit people. It's work that her grandmother Bertha Gunn always wanted to do: "No one would listen to me," she told her granddaughter, "but they're listening to you."

Polls show that a majority of Americans support a Green New Deal, but skeptics claim that it is too costly and promotes a "socialist" agenda by expanding government involvement in the economy. The plan calls for "meeting 100 percent of the power demand in the United States through clean, renewable, and zero-emission energy sources," removing "pollution and greenhouse gas emissions from the transportation sector as much as is technologically feasible," and guaranteeing jobs "with a family-sustaining wage, adequate family and medical leave, paid vacations, and retirement security to all people of the United States."

The goals of the Green New Deal are big. Ocasio-Cortez says, "This is going to be the Great Society, the moon shot, the civil rights movement of our generation."

A Practical Approach

Moon shots are sometimes the best way to get big things done. But some climate change activists like Ayana Elizabeth Johnson believe that it's time to be practical, which means prioritizing.

Johnson was instrumental in the development of the Blue New Deal, another name for an Ocean Climate Action plan, developed by a team that included Johnson in partnership with Senator Elizabeth Warren of the coastal state of Massachusetts. It focused on protecting, restoring, and renewing ocean habitats. As more than 70 percent of our planet is water, proponents of the Blue New Deal believe that our oceans are also a good source of "clean" energy and potential for economic growth. Johnson founded Ocean Collectiv, an organization that brought together a dream team of scientists, artists, lawyers, technologists, and others to work on science-based, community-driven ocean conservation strategies. She's now one of the leaders of Urban Ocean Lab, which focuses on climate policy in coastal cities, with an eye toward effecting larger-scale change.

Johnson believes that as the U.S. government continues to drag its feet, damage and destruction progresses at a rapid pace, so we need to accept that it's too late to save everything and go from there. She points to the coral reefs of the Florida Keys and U.S. Virgin Islands as an example of a high priority. "Overfishing, pollution and coastal development must be dramatically reduced" for them to survive rising sea levels. "For areas we can no longer maintain, we must make the most difficult of choices—give up, and accept that change is not always preventable . . . This is not to say we should de-list endangered species or revoke area protections," she goes on, but we should be honest when the costs outweigh the benefits of sustaining a particular ecosystem, or of helping a species survive in a certain habitat.

Acid in Our Oceans

Carbon dioxide naturally dissolves in ocean water, the two combining to make a weak acid that all sea creatures have evolved to live with. But as we release more carbon dioxide into the atmosphere, more dissolves in the oceans, making them more acidic. Some scientists call this acidification the "evil twin" of climate change. It's another consequence of too much CO_2 in our atmosphere.

Tiny changes in acidity make a big difference to sea creatures. Shellfish, for example, make their shells with the carbonate that occurs naturally in seawater. More acidity means less carbonate in the water, which means shellfish can't make their shells as well. If the acidity gets too high, the shells actually dissolve. Without decent shells, shellfish tend to die off and disappear, threatening the food supply of all the species that eat them, all the way up the food chain to humans. More acidification will affect fishing and fish farming, as well as tourism.

The same thing happens with coral reefs, which tiny organisms build over time with the carbonate in seawater. Increased acidity means the corals can't build and maintain the reef as well, and too much acidity will dissolve the reef. In addition, if temperatures get too high, corals bleach, which means they expel the algae that give them color. Algae are necessary for corals to live, so if corals are heat stressed for very long, they die.

Why is this important? Coral reefs protect our shorelines, providing a buffer against waves, storms, and floods and preventing erosion of beaches. The reefs provide habitat, nourishment, and medicine for a number of creatures. These underwater ecosystems are a major source of employment and food around the world.

FROM PENGUINS TO PEOPLE

For more than twenty years, researchers from a nonprofit group called Oceanites have been studying penguins in Antarctica, which is warming faster than anywhere else in the world except the Arctic. By analyzing the effects of climate change on Antarctic Peninsula penguins, scientists can understand how the rapidly changing climate will affect human beings, and why. Penguins have existed for sixty million years; they can teach humankind how to adapt to a changing climate. The Oceanites' work so far has shown that Adélie and chinstrap penguins are in significant decline, while gentoo penguins seem to be adapting well.

Plastic Debris and Our Seas

Our oceans play an important role in managing our climate and absorbing the carbon dioxide that we put in the atmosphere. We put things in the oceans that affect the climate as well: Plastics, which basically last forever, are a product of fossil fuels and emit greenhouse gases from their inception.

"The avalanche of plastics flowing into our oceans—equivalent to a dump truck-load every minute—is just the tip of the iceberg," says Jacqueline Savitz, a policy officer for the advocacy group Oceana. "On top of the choking sea turtles, starving seabirds and dying whales, we can add plastic-driven melting ice caps, a rising sea level and devastating storms. Whether you are a coastal resident or a farmer, a marine mammal or a sea turtle, plastic is the enemy."

In 2008, eighteen-year-old Kristal Ambrose was working at a Bahamas aquarium, treating a sea turtle that had ingested plastic. "For two and a half days . . . we pulled out one piece of plastic after another. My role was to hold down her front flippers. Now, sea turtles have salt glands to protect their eyes from the sand, so it looks as if they're crying. So while I'm holding down her fins, she's crying and I'm crying. I'm saying to myself, 'I'll never drop a piece of plastic on the ground again.'"

A few years later, Ambrose was on a ship in the middle of the Pacific Ocean, studying the ocean currents that trap debris, when she saw a giant patch of garbage made mostly of plastic. Horrified, she went home and started the Bahamas Plastic Movement to educate her fellow Bahamians about the destructive nature of plastic pollution. She involved kids in citizen science projects through workshops and a summer camp. At the end of one camp, the kids held a "trashion" fashion show, where the campers presented jewelry and purses made from plastic debris collected from the beach. The campers in 2017 drafted a bill and lobbied the Bahamian government to ban single-use plastic bags. They planned, they researched, they even sang to a government minister to get his attention. And it worked. The Bahamas phased out all single-use plastics by January 2020 and promised to join the UN Clean Seas campaign, through which countries commit to reduce or eliminate single-use plastics. Kid power strikes again!

Ambrose went on to become the first Bahamian winner of the Goldman Environmental Prize in 2020. "Know that your voice has weight. Your voice has power. You can do anything, and know that people are watching you, even when you think you're not doing your best," she said in her virtual acceptance speech, filmed at the beach during the COVID-19 pandemic with kids from her Plastic Debris and Me workshop playing in the ocean.

Doomsday Prep

Surrounded and partially hidden by snow, a slim, gray box of a building burrows deep into a mountain on an island between Norway and the North Pole. It can stand strong against floods, earthquakes, and is even designed to be bombproof. It looks like something out of a Hollywood movie, where villains could be plotting to destroy us all. It's actually nothing so sinister, quite the opposite. It's the Crop Trust's Global Seed Vault, an "insurance policy for the world's food supply." In other words, "the final back up."

The cold-storage facility in Svalbard houses all kinds of seeds for food crops, preserving their genetic data for the future. Plant breeders and researchers depend on seed banks around the world to develop crops with certain traits, such as better heat or cold resistance or more fruit production. This will be even more important in the face of climate change, which will make growing conditions vastly different. Should a natural or human-made disaster wipe out an institution's seed bank, a backup copy will exist in Svalbard. As the Crop Trust says, "The loss of a crop variety is as irreversible as the extinction of a dinosaur, animal or any form of life."

Opened in 2008, the Global Seed Vault today stores over a million varieties of seeds, and the Crop Trust is working to preserve even more, ensuring genetic diversity. The vault has the capacity to hold up to 4.5 million varieties, each sample containing five hundred seeds. That adds up to 2.5 billion tiny treasures. Regional seed banks deposit the seeds, which are stored in foil packets in sealed boxes on shelves, and only depositors can open their boxes and take out the seeds. It works just like the vault at your local bank—that is, if your bank was underground and kept constantly at −18 degrees Celsius (−0.4 Fahrenheit).

Back in 2015, scientists at a Middle Eastern seed bank were forced to flee the violence of civil war in Aleppo, Syria. To restore its important collection of seeds that grow in arid regions, the seed bank became the first organization to make a withdrawal from the Global Seed Vault. It has since deposited new seeds in Svalbard for safekeeping.

STUFF OF SCIENCE FICTION?

Animals like the Antarctic penguins can tell us a lot about the effects of climate change, but some scientists think animals in the Arctic could help us slow climate change. Genetic researchers at Harvard believe that bringing back the woolly mammoth could be vital in preventing further harm from greenhouse gas emissions. Arctic lands are covered by permafrost, areas that have been frozen for a million years, since the Pleistocene epoch. Permafrost contains huge amounts of carbon, and as Earth's temperatures rise, those areas will thaw and release the carbon into our atmosphere—raising the Earth's temperatures higher and faster.

In the very old days, scientists think, roaming mammoths may have helped keep the permafrost frozen. Like giant landscaping machines, mammoths would have pushed over trees and tamped down soil and winter snow on the tundra. The short grasses and mosses that thrived in this environment reflect more sunlight than trees, and compacted snow can't insulate as well, so scientists think the mammoths' presence kept the ground cooler.

Research today suggests as much. At Pleistocene Park on the Siberian tundra, scientists have been studying the effects on the landscape of free-roaming animals such as bison, horses, and reindeer. Early findings show that all their heavy tramping, eating, and pooping is changing the landscape for the better and even cooling the ground. Within the next decade, the Harvard geneticists hope to add mammoths to the mix at the park. They are trying to resurrect the mammoth by cloning living elephant cells that have been altered to include a bit of synthesized mammoth DNA.

The genetic diversity preserved in the Global Seed Vault often comes from seeds cultivated over generations by Indigenous farmers. For thousands of years, since the beginning of farming, Indigenous people have been collecting and exchanging seeds, developing genetic diversity as they grew hundreds of varieties of a crop in different locations and conditions.

"Scientists would just take seeds from us, not recognizing our knowledge," said Lino Mamani, an Indigenous potato farmer in the Andes Mountains in Peru. The scientists wouldn't share information or credit the success of the farmers' traditional methods.

But as climate change threatens everyone's food supply, scientists and traditional farmers are seeing the benefits of working together. As head of the Papa Arariwa (Potato Guardians) in Peru, Mamani sent a collection of potato seeds to Svalbard. "Climate change will mean that traditional methods of maintaining this collection can no longer provide absolute guarantees," he said. "The Vault was built for the global community and we are going to use it."

Potatoes are the most important noncereal food crop in the world, and they have been eaten in the Andes for eight thousand years. In the face of climate change, "it's time traditional knowledge and science work together," Mamani said.

Scientists today are taking the time to understand traditional methods and to help Indigenous farmers exchange knowledge with others from around the world. A while ago, scientists brought Indigenous potato farmers from China and Bhutan to Peru's Parque de la Papa (Potato Park). They exchanged ideas on how to cope with a changing climate that has made it harder for them to grow food on land they had farmed for generations. As Mamani said, "We can learn more from others with similar problems about technology that might be useful."

Chapter 14
THE CLIMATE JUSTICE MOVEMENT

Think Local—and Global

The environmental justice movement that focused on imbalances in local health hazards, waste management, and exposure to toxins is the foundation of today's climate justice movement. While a primary concern of the climate justice movement is exposure to greenhouse gas emissions, activists, researchers, and many scientists also believe that to address the crisis, we must transform our larger economic and political systems with an eye toward justice. "Climate action can't just be aimed at reducing emissions. It must focus on reorganising society to bring about a world that's grounded in justice and equity," says Andrea Ixchíu, a journalist and activist in Guatemala.

There are signs that governments are taking notice and taking action. In 2007, the U.S. Supreme Court ruled that the EPA could lawfully regulate carbon dioxide under the Clean Air Act.

In 2013, President Obama released his Climate Action Plan, offering three strategies of attack:

- Cut carbon pollution
- Prepare the United States for the impacts of climate change
- Lead international efforts to combat global climate change

Obama bypassed a gridlocked Congress and used executive actions to put his climate policy into place, but then the administration of President Donald Trump used the same strategy of executive action to cancel Obama's policy and other climate initiatives.

The Paris Agreement

In 2015, the United States joined almost two hundred countries around the world to sign on to the Paris Agreement developed at the twenty-first Conference of the Parties of the United Nations Framework Convention on Climate Change (the UNFCCC to those in the know) to address the crisis. Signatory nations agreed to work on reducing greenhouse gas emissions to decrease climate pollution and limit the rise in global average temperatures. And then in 2017, the Trump administration announced that it would withdraw in three years, which was as soon as legally possible.

As the U.S. federal government seemed to be leading a charge backward, communities that were thinking about global health were having to act even more locally. In 2019, New York City passed major legislation to address the climate crisis, putting a limit on climate-changing pollution from big buildings generated by electricity use, heating, and cooling, and requiring unprecedented cuts to greenhouse gases. According to the City Council, "Reducing emissions from our buildings is the most significant action the city can take to reduce greenhouse gas emissions in NYC, as buildings contribute nearly three-quarters of all citywide emissions."

This is a big deal for the second biggest city in North America because the United Nations reports that "cities will be home to an estimated two-thirds of the world's population by 2050, and already account for roughly three-quarters of greenhouse gas emissions."

On his first day in office, President Biden signed an executive order to re-enter the Paris Agreement, and in February 2021, the U.S. officially rejoined the accord.

In August 2021, the UN Intergovernmental Panel on Climate Change (IPCC) issued a report that warned, once again, of the dire situation we're facing: almost irreversible sea-level rise, and more deadly heat waves, droughts, and storms. "I think the message of the report is complex," says climate scientist Kate Marvel. She points out that the report is definitive about the human impact on the climate crisis, that scientific advancements mean that we can quickly assess the impact of climate change on specific events, and that while we have a long and difficult way to go . . . we've also come far. "You can't procrastinate for as long as we have. But it also says, look, the climate cares about how much carbon dioxide is in the atmosphere and you can be part of the solution. You can be part of the enormous group of people who are going to need to turn this thing around."

And three months later, when the twenty-sixth UN Climate Change Conference, called COP26, came around in Glasgow, Scotland, it was hailed by COP President Alok Sharma as promising "to be the most inclusive COP ever," with a spokesperson proclaiming that ensuring "the voices of those most affected by climate change are heard is a priority for the COP26 Presidency."

However, COVID-19 and immigration restrictions, a lack of affordable accommodations, and what many called poor organization led to continued marginalization of the most vulnerable, and to protests by farmers, youth, representatives of the Global South, Indigenous activists, and more.

"COP26 is a performance," said the Indigenous activist Ta'Kaiya Blaney of the Tla A'min Nation at the People's Plenary, a meeting of a coalition of grassroots activists. "It is an illusion constructed to save the capitalist economy rooted in resource extraction and colonialism. I didn't come here to fix the agenda—I came here to disrupt it."

The organization Global Witness noted that representatives from the fossil fuel industry were actually the largest delegation at the conference. "At least 503 fossil fuel lobbyists, affiliated with some of the world's biggest polluting oil and gas giants, have been granted access to COP26, flooding the Glasgow conference with corporate influence." Global Witness's Murray Worthy added in a statement, "The presence of hundreds of those being paid to push the toxic interests of polluting fossil fuel companies, will only increase the skepticism of climate activists who see these talks as more evidence of global leaders' dithering and delaying."

Some pointed to what they saw as real progress at COP26. As vulnerable nations pressed richer and more culpable countries to pull their weight in the fight against this crisis, the U.S. and China announced an agreement to spend the next decade collaborating to reduce greenhouse gas emissions. U.S. Climate Envoy John Kerry said that the "two largest economies in the world have agreed to work together on emissions in this decisive decade . . . This is a roadmap for our countries and future collaboration." Critics noted that the language of the agreement was vague and inadequate; the same criticisms were leveled as the conference attempted to draft a final agreement and path to a future that met the goals of the Paris Agreement and avoided calamity.

Reuters reported that the agreement, which "attempts to ensure the world will tackle global warming fast enough to stop it [from] becoming catastrophic, is a balancing act—trying to take in the demands of climate-vulnerable nations, the world's biggest polluters, and nations whose economies rely on fossil fuels." Activists from the UK's COP26 Coalition issued a document calling for an end to investing in fossil fuels, reparations and debt cancellation for more vulnerable countries, and more.

"I Can't Breathe": Coronavirus, Systemic Racism, and the Climate Crisis

On November 6, 2012, four years before he was elected president, Donald Trump tweeted, "The concept of global warming was created by and for the Chinese in order to make U.S. manufacturing non-competitive." The infamous tweet received 68,000 likes. Four years later, as he campaigned for president, he claimed this comment was a joke. He went on to say, "I think the climate change is just a very, very expensive form of tax. A lot of people are making a lot of money . . . And I often joke that this is done for the benefit of China. Obviously, I joke. But this is done for the benefit of China, because China does not do anything to help climate change. They burn everything you could burn; they couldn't care less." Asked during his presidency if he still believes climate change is a hoax, he said in a television interview, "Something's changing, and it'll change back again. I don't think it's a hoax." He added, "But I don't know that it's manmade." As Trump consistently undermined the scientific consensus that humans had contributed toward warming, many officials in his administration parroted his words in their own comments and actions.

Americans have seen time and time again the destruction caused by slow and inadequate government response to what science says—what the facts are, right in front of us. The public health impact of environmental injustice is devastating, and hurts the most vulnerable, who are doing the least damage. The link between racial and climate justice was strikingly visible in 2020 as people

took to the streets after multiple incidents of police killings of Black people, speaking out from behind face coverings and masks in an attempt to protect themselves from COVID-19, the disease caused by the global coronavirus pandemic. COVID-19 tragically amplified all that we already knew. In February 2020, President Trump said that COVID-19 would somehow just "disappear." He often insisted on calling it the "Chinese virus" and "the kung flu," and when it was pointed out to him that such language was fuel for racism and violence against Asians and Asian Americans, claimed that he "wanted to be accurate."

But wait, this is a disease, right? A sickness doesn't discriminate! In fact, some called COVID-19 a "great equalizer." Once again, that's not how things are turning out. By the time Trump left office in January 2021, half a million Americans had died from COVID-19. In the United States and other parts of the world, the deadly virus hit Black, Latinx, and lower income communities the hardest. The Centers for Disease Control found that nearly two times as many Black people were dying of COVID-19 than white people. Native American and Latinx patients were dying at an even higher rate. Why? As one Black observer puts it, "Because we're far more likely to live in food deserts, and near dumping grounds, power plants and large-scale animal farms, all of which saddle us with preexisting conditions like asthma, diabetes and heart disease. Those are the same preexisting conditions that authorities attempted to blame for the deaths of Eric Garner and George Floyd." (Food deserts are neighborhoods where there's very little fresh produce for sale.)

While Trump rejected the evidence of the public health impact of environmental injustice in favor of misinformation and racism, scientists found that higher levels of air pollution, such as in urban communities with higher concentrations of people of color and low-income residents, corresponded with greater vulnerability to COVID-19. The relaxing of air pollution regulations during the Trump administration had a harmful impact on the health of those in poor communities.

"Oil, gas, and petrochemical industries have concentrated so heavily in low-income, majority-black-and-brown areas that black people are 75 percent more likely to live near industrial facilities than the average American," writes Rhiana Gunn-Wright, the same policy analyst who worked on the Green New Deal. "The people most likely to die from toxic fumes are the same people most likely to die from Covid-19."

Even as research builds to show that this matter of life and death is linked in many ways to the climate crisis, the U.S. government repeats old patterns. Like much of the rest of the world, our country was forced to shut down as people quarantined to slow the spread of the disease. The economy suffered, businesses shuttered, jobs were lost. And in developing a plan to revitalize the economy, many politicians took the opportunity to strike . . . at climate justice policies and the Green New Deal, and to offer support to the fossil fuel industry.

Commentators warned that in the United States, the world's second-largest carbon emitter, any "consideration of a green recovery appears dead on arrival." Republicans warned that the Green New Deal would prevent U.S. citizens from getting the economic help and support they needed. President Trump sought to bail out the oil and gas companies.

Observers like Rhiana Gunn-Wright saw reason to be "furious": Of the nearly $2 trillion in aid proposed in the first version of the relief bill, $500 billion went toward a business-relief fund with little to no oversight, she notes, including $58 billion for airlines, which contribute more than 10 percent of U.S. transportation emissions of greenhouse gases. "A lax definition of eligible businesses created a loophole for oil and gas. The bill included no climate protections, so the claim that it was being held up over Green New Deal provisions was absurd."

Here we go again. Even though it's possible that health crises like COVID-19 might become more frequent and widespread as a result of our changing weather patterns and continued reliance on and support of the fossil fuel industry, the Trump administration and U.S. leaders relied on hollow speeches, misdirection, and baby steps. "Instead of leading this transition," said a former climate adviser to President Obama, "we're trying to impede the inevitable as others move quickly forward."

TROUBLED WATERS

At eight years old, Mari Copeny, aka Little Miss Flint, understood the connection between equity and the environment all too well. In 2016, she wrote to President Obama about the chronically contaminated water in Flint, Michigan, inviting him to visit her and see what was happening for himself.

State-appointed city managers (not elected by city residents) had decided to save some money and switched the city's water supply to the Flint River in 2014. Residents immediately reported a change in the water's color and taste. Within months bacteria forced residents to boil water. Then toxic chemicals were found. Finally, repeated scientific tests showed dangerous levels of lead; corroding pipes were leaching it into the water. Although officials continued to insist the water was safe to drink, Flint children had alarmingly high levels of lead in their bloodstream. This can cause a number of health issues, including learning difficulties, hearing problems, delayed puberty, and behavioral disorders.

Obama answered her letter and visited her in Flint, shining a bright light on the crisis.

Because of discriminatory housing policies, Black and Brown people have been forced to live in places like Flint or in "asthma alleys" like the Mott Haven area of the Bronx, where residents inhale the toxic air generated by four highways. A 2019 article in *Proceedings of the National Academy of Sciences* put it this way: "Racial–ethnic disparities in pollution exposure and in consumption of goods and services in the United States are well documented . . . Black and Hispanic minorities bear a disproportionate burden from the air pollution caused mainly by non-Hispanic whites."

And during that period when many were quarantined, "shut down," and living in a state of "pause," one thing did move forward—a period of recovery and renewal of the Earth's natural resources. The cloudy canals in Venice, Italy, cleared up. Air pollution in China dropped dramatically. In San Francisco, New York, and Seattle, the levels of dust, dirt, and soot in the air lowered considerably after a short period of stay-at-home orders. But these positive effects turned out to be only temporary. By March 2021, with the world returning to business as usual, pollution and carbon emissions were returning to their high levels as well.

On January 20, 2021, President Biden returned the United

Chapter 15
TOMORROW PEOPLE

Tomorrow belongs only to the people who prepare for it today.
—**Malcolm X,** Muslim minister, Black Nationalist, and human rights leader

Teach Them Well and Let Them Lead the Way

In some ways, the climate movement can replicate the patterns of marginalization that we see in the larger culture. "The truth is the same throwaway culture that disposes our planet disposes of people too—especially people of color," writes Wanjiku Gatheru, founder of Black Girl Environmentalist.

Writer and activist Mary Annaïse Heglar has pointed out that the climate movement is often perceived as dismissive of the very people on the frontlines of the crisis. "It's not just time to talk about climate," she says. "It's time to talk about it as the Black issue it is. It's time to stop whitewashing it. In other words, it's time to stop #AllLivesMattering the climate crisis."

Black people and other marginalized groups have long been unwelcome in U.S. public green spaces, particularly through legal Jim Crow segregation, then the lingering impact of that injustice.

Despite this, the story of the close relationship between Black people and the American environment is rich and deep, "from the time of slavery to the farming cooperatives of the mid-20th century, the preservation of traditional food production, and even healing root work in the South." Black, Indigenous, and many other people of color have been involved in the environmental movement since there has been a movement.

Today, organizations such as Outdoor Afro, Outdoor Asian, Native Women's Wilderness, and Latino Outdoors work to build on that history. "¡Estamos Aqui! / We Are Here!" is the message of community and outdoor engagement organization Latino Outdoors. "Public lands belong to everyone," founder José González says. "All communities deserve access to our natural resources and the subsequent health and economic benefits." And the expertise of those who are directly experiencing the impact of the climate crisis seems vital to the movement's sustainability and success.

You can be a kid, you can be a grandmama, an abuela, a professional person—anybody can do it. That makes it messy and complicated but that's also what makes really beautiful: anybody can get involved to preserve their natural and cultural environment.

—Leslie Fields, Sierra Club, Director of Policy and Advocacy

Jacqueline Patterson, director of the NAACP Environmental and Climate Justice Program, also believes in the power of an inclusive movement. "I am inspired by some of the work that's happening at the local level connecting various social justice issues with environmental issues. The NAACP . . . worked with the Department of Corrections to do training in solar installation for folks while they were incarcerated. That resulted in people having skills to be placed in jobs as they came out of incarceration."

Isra Hirsi is also thinking of transforming the world. At sixteen, she was one of the founders of the U.S. Youth Climate Strike. "Getting young people out, going to state capitols, going to city halls, going to the nation's capital and talking about these things, that says something," she believes. "That's what we're trying to do: Change the conversation, not only about things like the Green New Deal but so much more."

It's not surprising that she thinks big and boldly about making a difference. Isra Hirsi is the daughter of Ilhan Omar, a representative from Minnesota, the first Somali American and one of the first two Muslim women elected to Congress. In the lead-up to the 2020 presidential election, Hirsi spoke out to encourage Democratic candidates to debate one another on climate issues, and CNN ended up hosting that debate.

Hirsi is inspired by the ways that women of color and people from other marginalized communities are at the forefront of climate action, and believes that building an inclusive movement that empowers all voices is vital. "It's important to talk about what climate change does to marginalized communities . . . Environmental racism is a really big thing."

Hirsi believes that people who want to get involved in the environmental movement should join "inclusive groups or diverse groups led by people of color or women of color . . . put those people at the forefront of their own movement."

Today, young people continue to be some of the loudest, most creative and committed voices at the forefront of the climate crisis conversation. Kids all across the world know, just as Whitney Houston sang, children are our future. Literally. Even if adults have not taught them well, young people are ready to lead the way. That means making sure conversations about how we interact with our environment are a regular part of school life.

And the climate conversation is not just about health and recycling. "For us, it's about our islands sinking. Our culture—all of it— would go away," eighteen-year-old Pone Aisea of Portland, Oregon, told the *New York Times*. Aisea's family is from the island of Tonga. Many families in her community have roots in places most vulnerable to the effects of the climate crisis. So she and the students at her high school formed the Pacific Islander Club and in 2016 got the school board to adopt a resolution that called for including the climate crisis in the curriculum of Portland public schools.

Swedish teen Greta Thunberg is distantly related to Svante Arrhenius (remember him, the scientist who predicted in 1896 that fossil fuel use would warm the planet?) and first heard about climate change when she was about eight years old. She had questions. "I remember thinking that it was very strange that humans, who are an animal species among others, could be capable of changing the Earth's climate," she said in her 2018 TEDx Talk in Stockholm. "Because if we were, and if it was really happening, we wouldn't be talking about anything else."

But to Thunberg's surprise, adults seemed remarkably unbothered. "If burning fossil fuels was so bad that it threatened our very existence, how could we just continue like before? Why were there no restrictions? Why wasn't it made illegal?"

The lack of action on something that appeared to be a crisis affected Thunberg deeply. Thunberg, who is autistic, once said that "For those of us on the spectrum, almost everything is black or white. We aren't very good at lying." So it was distressing to be in a world that was lying to itself. "Everyone keeps saying climate change is an existential threat and the most important issue of all, and yet they just carry on like before. I don't understand that, because if the emissions have to stop, then we must stop the emissions . . . There are no gray areas when it comes to survival. Either we go on as a civilization or we don't. We have to change."

Inspired by the teens who walked out of class in 2018 to protest gun violence in the United States, Thunberg skipped her own classes and started protesting outside the Swedish parliament. She

stayed out there every school day for three weeks, more protestors joining her every day. After the Swedish elections she went back to class, mostly. She and her fellow activists continued to strike every Friday, starting a worldwide movement, Fridays for Future. By 2021, Fridays for Future had recorded ninety-seven thousand climate strikes around the world, with fourteen million strikers.

In August 2019 Thunberg sailed across the Atlantic on an emissions-free racing yacht to New York City to address the UN Climate Action Summit. "How dare you!" she told world leaders in a speech that quickly reverberated across the globe. "You have stolen my dreams and my childhood with your empty words."

During her visit to the United States, Thunberg met climate activist Tokata Iron Eyes, who had spoken out at Standing Rock four years earlier, and began her environmental activism at the age of nine, when she protested against uranium mining in South Dakota's Black Hills. Iron Eyes invited Thunberg to visit her people's homelands in the Dakotas. The two teenagers spoke at the Pine Ridge and Standing Rock reservations in South Dakota and led a climate rally in Rapid City.

"We know what we need as individuals and we know what makes us happy, what makes us healthy," said Iron Eyes at Pine Ridge. "And the fact that we can recognize that and deny that to the people around us by being complacent is really a strange phenomenon." At the Rapid City rally the following day Iron Eyes stressed: "This crisis does not care about the imaginary political boundaries that we put up among each other. This crisis does not care whether you're rich or poor, you're Black, you're white or Indigenous."

And to the five hundred students gathered at Standing Rock High School, she said that the world is "at the edge of a cliff as to how much time we have to save our communities." In reference to Thunberg, she added: "No 16-year-old should have to travel the world in the first place sharing a message about having something as simple as clean water and fresh air to breathe."

Making It Count

Young people wanting to get involved in the climate movement can look to the examples of other young climate activists for advice and encouragement.

Follow your own path: Mexican-Chilean activist Xiye Bastida received the Spirit of the United Nations Award as a high schooler in 2018 for her leadership as a youth organizer on climate issues, and hasn't looked back. As a member of the Otomi Toltec Nation living in New York, Bastida "noticed that my classmates were talking about recycling and watching movies about the ocean. It was a view of environmentalism that was so catered toward an ineffective way of climate activism, one that blames the consumer for the climate crisis and preaches that temperatures are going up because we forgot to bring a reusable bag to the store." She went on to major in Environmental Studies in college, and her work is focused on our relationship to the Earth and one another, and rooted in the Indigenous culture in which she was raised.

Make sure you are heard (and seen): Vanessa Nakate, who led an effort to preserve the rainforests of the Congo, refuses to be erased, even when media coverage tries to do exactly that. The Associated Press cropped the founder of the Rise Up Movement and Youth for Future Africa out of a photo with white climate activists. The AP is a major news agency whose stories and photos are used by thousands of news outlets around the world. Nakate spoke out, and the AP issued an apology. "As much as this incident has hurt me personally, I'm glad because it has brought more attention to activists in Africa," she said. "Maybe media will start paying attention to us not just when we're the victims of climate tragedies." In 2019, Nakate started the Green Schools Project to supply solar energy panels to Ugandan schools, and she continues to champion the voices of the marginalized in the movement and prioritize community-oriented solutions. "I have been talking to my fellow activists about bringing social solutions, not just technological solutions, into climate activism."

In her memoir, *A Bigger Picture: My Fight to Bring a New African Voice to the Climate Crisis*, Vanessa Nakate writes about conversations and collaborations with other African climate justice activists like Kaossara Sani (Togo), Adenike Oladosu (Nigeria), and Elizabeth Wathuti (Kenya). "I see my role in climate activism as bringing up conversations that many people have never had, and highlighting the destructive policies and investments of banks, hedge funds, multi-national corporations and governments—all of which would like the rest of us to have no idea what they're up to . . . What happens in the Congo Basin rain forest doesn't just affect people in countries in central Africa; it influences weather patterns across the world. The climate crisis respects no geopolitical borders, political bloc or regional trade associations. So what happens in the Congo isn't just the business of the Congolese or their neighbors. It concerns all of us."

Think global, act local: After watching Al Gore's *An Inconvenient Truth* at sixteen, Chinese student Ou Hongyi stopped going to school. Alarmed by the climate crisis, she felt activism was more important. In an authoritarian country like communist China, where protest is viewed with suspicion if not worse, Ou chose to take a stand—and has been ostracized and ignored, as well as harassed by authorities. She has held solitary Fridays for Future climate strikes in her hometown of Guilin and staged an overnight vigil outside a Guangzhou hotel. "All of us know that in the hotel industry, the bedding for guests and other disposable items the hotel provides waste a lot of water resources and emit a lot of carbon dioxide," she said. She also started a tree-planting initiative. As China battles air and water pollution, flooding, and its status as the world's largest emitter of carbon dioxide, Ou shares her message on social media (where she goes by Howey Ou). "China doesn't need one more climate scientist—there are so many and all of them say the science is clear. What China does need is one more climate activist to push for change, for action from the government and from the public."

Educate yourself so you can educate others: "What does climate change have to do with human rights?" Vic Barrett wondered when his after-school human rights program took on the crisis. But as he learned more about the impact of weather events like Hurricane Sandy on people like his own Garifuna Honduran family, he made the connection. Barrett went on to become one of the twenty-one plaintiffs in the constitutional lawsuit *Juliana v. United States*, which put a spotlight on youth climate activism. Barrett's intersecting identities—Afro-Latinx, Indigenous, transgender, first-generation American—inform his activism on environmental justice. He recommends reading *Parable of the Sower* by Octavia Butler. "I think imagining the future is a really important part of being a climate activist," he says, "and being able to design the future that you want and get imaginative with it. And just in general, doing a lot of educational work, doing work to unpack your own biases." Barrett has spoken at the United Nations and was a leader of the Global Climate Strike in New York in September 2019.

Lead by example: Growing up in the Wiikwemkoong First Nation along the shores of Lake Huron, Autumn Peltier learned early on that many Indigenous communities in Canada had not been able to drink their water for years because of contamination. She made headlines as a twelve-year-old when she confronted the Canadian prime minister about it, and at fourteen, she was named the chief "water protector" for the Anishinabek Nation, taking over from her great-aunt, Josephine Mandamin. "I advocate for water because we all came from water, and water is literally the only reason we are here today and living on this earth," Peltier has said. The COVID-19 pandemic heightened the critical need to clean up the water supply of Indigenous communities. "Clean water is needed not only for drinking, but for washing hands, brushing teeth, doing dishes and cooking," she said. "That's simple sanitation. Many First Nations communities do not have access to simple sanitation." An International Children's Peace Prize nominee, Peltier has spoken at the United Nations and was the subject of the 2020 short documentary *The Water Walker*.

You never know how far a new idea will go: Having already started her own business at age eight, Maya Penn founded Maya's Ideas 4 the Planet when she was eleven. Her nonprofit organization has provided eco-friendly sanitary products to girls who can't go to school because of their monthly cycle and masks made from recycled material to health care workers during the pandemic. Today her "slow fashion" brand, Maya's Ideas, sells artisan-made clothing and accessories created from organic, recycled, and vintage materials. "I feel that I am part of the new wave of entrepreneurs that not only seeks to have a successful business but also a sustainable future," she said. "I believe that I can meet the needs of my customers without compromising the ability of future generations to live in a greener tomorrow by being an environmentally responsible individual and business owner." Her ideas continue to grow: She has written a book; advises big companies on sustainable practices; and has become an animator, producing films with environmental and social themes.

Young people are at the forefront of holding the more powerful accountable, highlighting instances of "greenwashing" (using the environmental movement as a marketing tool), and "woke washing," which *Teen Vogue* writer Laura Pitcher describes as "where ethically problematic companies use social movements to increase sales without addressing how their business is complicit."

However, they are clear that this situation should not be dumped on them to "fix." Sixteen-year-old activist Alexandria Villaseñor warned on Twitter: "Let's cancel the Youth Climate Hope Industrial Complex now . . . If you're relying on youth to save us, then you're not doing enough yourself. Get in the streets, call your lawmakers, phone bank, talk to your friends, families & communities. We're not going to save you."

Kids' Day in Court

In 2015, twenty-one young people in the United States filed the landmark lawsuit *Juliana v. United States* against their government, asserting that it "has violated the youngest generation's constitutional rights to life, liberty, and property, as well as failed to protect essential public trust resources." The fossil fuel industry joined the government's side and tried to get the case dismissed, but their motion was denied.

"When those in power stand alongside the very industries that threaten the future of my generation instead of standing with the people, it is a reminder that they are not our leaders," Xiuhtezcatl Roske-Martinez, one of the suit's young plaintiffs, said at the time. "The real leaders are the twenty youth standing with me in court to demand justice for my generation and justice for all youth. We will not be silent, we will not go unnoticed, and we are ready to stand to protect everything our 'leaders' have failed to fight for. They are afraid of the power we have to create change. And this change we are creating will go down in history."

After a protracted battle to get the suit dismissed, a divided three-judge panel of the Ninth Circuit Court of Appeals finally ruled in the government's favor, saying in January 2020 that the federal courts could not provide a remedy for the young plaintiffs: "The plaintiffs' case must be made to the political branches or to the electorate at large, the latter of which can change the composition of the political branches through the ballot box."

The *Juliana* plaintiffs requested a rehearing with all eleven judges of the Ninth Circuit, but in February 2021, the full Ninth Circuit upheld the panel's ruling and dismissed the case. A federal judge ordered both sides to enter into settlement negotiations, but after five months, an agreement could not be reached, and the young people hoped to be allowed to take their case to trial. In

November of 2021, just before the start of the UN's COP26 conference on climate change, Julia Olson, lawyer for the young people said, "They have consistently said, in the Juliana case, there is no right to a climate system that sustains life, that these children don't have that right. And that's the position that we're seeing on the international scale too right now."

Chapter 16

GIVE LIGHT, AND PEOPLE
WILL FIND A WAY

~~~~~~

It's Not Easy Being Green

This is . . . a lot.

It is a lot to learn; it is even harder to learn how to live with the knowledge at a time with such big, frightening challenges. Survivors of disasters like Hurricane Maria in 2017 experience post-traumatic stress disorder; as many as one in fourteen children in Puerto Rico were afflicted, according to a recent study. Inuit people in Canada are increasingly anxious about their lives as temperatures rise and the frozen rivers and streams they rely on to reach hunting and fishing grounds begin to disappear. If you're frightened or anxious or depressed by the climate crisis and how we've handled it, you are not alone.

As *Losing Earth* author Nathaniel Rich puts it, "How does a sentient person alive now—the world already having warmed by more than 1 degree Celsius, with another 0.5-degree warming inevitable, and emissions continuing to rise unabated—how does one live with the knowledge that the future will be far less hospitable than the present? Should we obsess over it, ignore it, find some tense middle territory? What do our failures say about our substance as a people, as a society, as a democracy? Will future generations be satisfied with the answers we offer for inaction? If we vote correctly, eat vegan, and commute by bicycle, are we excused the occasional airplane ticket, the laptop, the elevators, year-round strawberries, trash collection, refrigerators, Wi-Fi, modern health care, and every other civilized activity that we take for granted? What is the appropriate calculus? How do we begin to make sense of our own complicity, however reluctant, in this nightmare?"

A National Wildlife Federation report on the psychological effects of climate change estimates that 200 million Americans will be exposed to serious psychological distress from climate-related events and incidents. "Some Americans already are or will soon experience anxiety about global warming and its effects on us, our loved ones, our ecosystems, and our lifestyles," it stated in 2012. "This anxiety will increase as reports of the gravity of our condition become more clear and stark."

And the thing is, it *should* feel that way, writes Mary Annaïse Heglar. "People can't fix the problems if they don't confront them. We have to let people mourn so they can come out on the other side. Perhaps it's time to stop worrying about giving people hope and to start letting people grieve." And with the situation as it continues to be, maybe we're not meant to simply dry our tears and "get over it" or get used to it. "As uncomfortable as those emotions are," continues Heglar, "I take comfort in the fact that I still experience them. I never want to be the person who can look at so much suffering in this world and feel fine. It hurts because it's supposed to. Turning off the pain would mean turning off my humanity, and I just can't do that."

Many scientists and activists report experiencing a deep sadness that can make it hard to continue doing their work. To address those feelings, one Rhode Island activist set up a table in a downtown plaza with the sign, CLIMATE ANXIETY COUNSELING 5¢. Like Lucy with Charlie Brown, she wasn't exactly a trained therapist. "A lot of what I do is listen and ask questions," she said. Scientist and conservationist Samantha Whitcraft persists in the face of grief by celebrating "the little wins."

Scientists have begun to formally study "ecological grief" or "climate grief," and some believe that acknowledging and explaining it can help human beings "reach across differences to connect with others" and are looking for strategies to help people deal with the pain that comes from "environmental loss." It's healthy to grieve, they remind us, saying that "grief can . . . strengthen and bring maturity."

The founders of the Good Grief Network suggest these steps to help people manage that grief in productive ways:

- Accept the severity of the predicament
- Acknowledge that I am part of the problem … and the solutions
- Practice living with uncertainty
- Practice gratitude
- Reinvest in meaningful efforts

Cofounder LaUra Schmidt says, "What helps people is building community, talking openly about the problem and how it affects them. There's a lot of pain about the climate people are bottling up." As Sarah Jaquette Ray writes, fear or even dread about the future has been a part of Black and Brown peoples' lives for a long, long time, "whether that terrain is racism or climate change. Climate change compounds existing structures of injustice, and those structures exacerbate climate change." She adds, "Instead of asking 'What can I do to stop feeling so anxious?', 'What can I do to save the planet?' and 'What hope is there?', people with privilege can be asking 'Who am I?' and 'How am I connected to all of this?'"

It's natural and normal to feel sad about all of this. There are strategies that can help you manage those emotions. "Doing something every day can be profoundly helpful in giving people hope, and a path to change," says graphic artist Sarah Lazarovic. "Tackle small things that add up to larger decisive, strategic goals. Don't think about the drops, focus on the bucket!" Two more bits of her practical advice:

- Choose one specific target for action. Instead of "big oil" choose one company to focus on.
- Share your reasons for climate crisis action without being judgy. "Be bold, but not a buzzkill."

MORE IDEAS TO COPE WITH CLIMATE CHANGE ANXIETY

Make mudpies! In "By Reconnecting with Soil, We Heal the Planet and Ourselves," farmer Leah Penniman writes that our ancestors had a healthy respect for soil and the importance of dirt in our lives. "The soil stewards of generations past recognized that healthy soil is not only imperative for our food security—it is also foundational for our cultural and emotional well-being." Modern scientists are now catching up, in experiments with mice that establish a connection between exposure to soil microbe *Mycobacterium vaccae* and the production of serotonin, the mood-regulating hormone, in their brains. In other words, rolling up your sleeves and playing in the dirt is a scientifically proven way of caring for your mental health.

Plant a tree to hug! A lot of carbon can be captured by one single tree. A European research team found that expanding forests, which will draw CO_2 out of the atmosphere, "could seriously make up for humans' toxic carbon emissions." The study shows that "the restoration of trees remains among the most effective strategies for climate change mitigation." According to the researchers, there is room on the planet for an additional 900 million hectares of tree cover, which could store over 200 metric gigatons of carbon. Getting involved with tree-planting initiatives such as Wangari Maathai's Green Belt Movement is a literal ground-level way of stemming the effects of carbon pollution.

Get with a group. Working in a community can be energizing and empowering by strengthening relationships with the Earth and with each other. There are all kinds of groups out there, and many ways to take action. "Groups are more effective than individuals," says professor of psychology and environmental studies Susan Clayton. "You can see real impact."

Become a citizen scientist. You don't need a white coat or a lab; you can get involved in environmental science projects right from your own home or community. Citizen science is simply recording your observations of the natural world, which helps scientists collect and analyze data over large areas and long periods. Institutions like the Cornell Lab of Ornithology and Zooniverse offer many opportunities to be a part of "people-powered research" related to climate and environmental issues.

There's a smorgasbord of opportunity to get involved. There are youth-oriented organizations like Sunrise Movement and Alliance for Climate Education and faith-based efforts like GreenFaith and Young Evangelicals for Climate Action. The NAACP Environmental and Climate Justice Program and Hip Hop Caucus highlight the work of people of color, and 350.org takes it to the streets with rallies, marches, and more. Activist Tesicca Truong got fired up about water waste at the age of six during a school assembly performance. Then she went on to build a campaign for sustainable water practices in her high school; cofound an organization, CityHive, promoting youth civic engagement; and act as a government adviser on environmental issues. When we see headlines like "Activists Remove 40 Tons of Plastic Waste from Pacific Ocean," we can grieve the conditions and actions that brought about those conditions, and be heartened by the people all over the world doing their part to bring change, like James Wakibia in Kenya, whose social media and letter writing campaign led to a 2017 ban on single-use plastic bags in that country.

GREEN FLAG SCHOOLS

The National Wildlife Federation launched the Eco-Schools USA program to help K–12 schools promote environmental education and sustainability practices. Students, administrators, staff, and community members work together to build a green school community. Schools that have done exemplary work in this area can win the Green Flag Award.

As has often happened in the history of social movements, artists are stepping up to illuminate the possibilities of a transformed society. The Dream Unfinished, an "activist orchestra" cofounded by Eun Lee, promotes dialogue on social and racial justice through its classical concerts. In 2019 it held a "Deep River" concert featuring works by composers who had been directly affected by the climate crisis and other environmental issues. Others, like Orchestra for the Earth, Beethoven Orchestra Bonn, named the first UN Climate Change Goodwill Ambassador, and the composer Einaudi, are also making music to inspire change. In hip-hop, Childish Gambino's laid-back sounding "Feels like Summer" includes lyrics that reference dangerous climate trends and their effect on the natural world.

In 2018, the Storm King Art Center in New York's Hudson Valley launched the exhibition *Indicators: Artists on Climate Change*. Artist Justin Brice Guariglia asked people to consider all the ways *We Are the Crisis* in his piece, a solar-powered highway sign, putting a spotlight on human impact on the climate. "Climate change isn't one thing with one way of looking at the world," said curator Nora Lawrence. "We wanted to look at climate change like a syndrome with different factors that taps into something larger, because not every artist is approaching it from the same angle."

In its *General Assembly* piece at Storm King, the art collective Dear Climate put up flags in the fashion of those at the United Nations, but with messages like "Say hello to hurricanes," "See the sea levels," and "Give me luxury or give me breath." Dear Climate uses posters like these "to cultivate a sense of affection for the

climate, and to recognize its ancient and complex relationship to human cultures," say the founders.

Visual artist Allison Janae Hamilton uses video and photographs to spotlight landscapes in the American South and focuses on "how the increasing vulnerability of the environment is directly related to that of certain communities that live within it, on how climate-related natural disasters illuminate existing social and political inequality."

Make It Count

Doing one small thing can make a difference for sure, but it can be tempting to stop there, like "Okay, I'm all about Meatless

Mondays! Climate crisis averted!" Carina Barnett-Loro, senior program manager at the Climate Advocacy Lab, points out that sometimes the sense of satisfaction from doing a good deed like recycling lessens the sense of urgency around the bigger issue. Barnett-Loro recommends *The Psychology of Climate Change Communication: A Guide for Scientists, Journalists, Educators, Political Aides, and the Interested Public* (okay, so, everyone). It's full of tips to help you keep the climate justice momentum going. "When people try to act in environmentally friendly ways, they often, in fact, do further harm to the environment," researcher Patrik Sörqvist found. They think that their little positive actions cancel out the big negative ones.

> We know of course there's really no such thing as the "voiceless." There are only the deliberately silenced, or the preferably unheard.
>
> **—Arundhati Roy,**
> author and human rights activist

Listen, learn, and listen again. Instead of shaming people for using plastic straws (and dismissing the many people with disabilities or illness who must use them), or ignoring cultural differences, traditions, and contexts that don't line up with your idea of what's sustainable, consider trying to amplify the voices and work of others, and keep your focus on what's best for all of us, in community. "This is a call for eco-social justice: acknowledging how factors such as racial injustice, colonialism, and capitalism have created social divides as well as environmental disasters like plastic waste," writes Nicci Attfield. Many believe that this vision of connectedness, and recognition of unjust systems, is at the foundation of successful large-scale transformation. Ayisha Siddiqa, founder of Polluters Out, sees love at the heart of that call. "The fight for climate justice is at its core a fight for love, it comes from a place of deep deep love for humanity." Many believe that this vision of connectedness, and recognition of unjust systems, is at the foundation of successful large scale transformation. The "civil rights movement transformed society in the United States because it was fundamentally rooted in a love ethic," writes bell hooks. "It is in choosing love, and beginning with love as the ethical foundation for politics, that we are best positioned to transform society in ways that enhance the collective good."

RETHINK RECYCLING

For decades, Western countries like the United States and Canada were "recycling" by sending large amounts of waste to countries in Asia. "Often, the shipments included contaminated waste that couldn't be recycled but made it past customs checks anyway, and countries had few legal avenues to send it back," reports the *Los Angeles Times*. Sometimes all kinds of materials—food, wood, plastic—were mixed together. Sometimes nonrecyclable plastics were dumped in with recyclables.

Then in 2017 China said, No more. It imposed a ban, and other countries, like Vietnam, followed suit, no longer accepting the richer countries' trash. "In the long term, the problem has to be solved at source. North America and Western Europe must take clear and conscious efforts to reduce waste," said Lawrence Loh, a professor at the National University of Singapore. "Rather than looking for the next place to dump waste, advanced countries should bear the responsibility of cutting on waste generation through sustainable practices."

After all we've learned, all we know, all that's been documented, that might sound impossible. It sounds good, but it's not easy to choose love. But here's the thing: We've done it before. Don't make the mistake of doing what Mary Annaïse Heglar calls divorcing "the environmental movement from a much bigger arc of history." Black, Brown, and Indigenous communities have resisted abusive systems and threats to their very existence for centuries.

When popular movements have managed to transform public opinion in a brief amount of time, forcing the passage of major legislation, they have done so on the strength of a moral claim that persuades enough voters to see the issue in human, rather than political, terms. If we speak about climate as only a political issue, it will suffer the fate of all political issues. If we speak about climate as only an economic issue, it will wither away like all moral crises reduced to penny pinching. The first requirement is to speak about the problem honestly: as a struggle for survival.

Everything is changing about the natural world, and everything must change about the way we conduct our lives. It is easy to complain that the problem is too vast, that each of us is too small. But there is one thing that each of us can do ourselves, in our own homes, at our own pace—something easier than taking out the recycling or turning down the thermostat, something more valuable. We can call the threats to our future what they are. We can take an honest look at who we were, who we are, and who we can be. We can ask: Whose voices are heard, and whose are suppressed? We can talk about the work of Robert Bullard, Wangari Maathai, and LaDonna Brave Bull Allard, the history and legacy of those who have been disregarded and discarded by those in power, and look to their examples of commitment to true liberty and justice for all.

I am convinced that if we are to get on the right side of the world revolution, we as a nation must undergo a radical revolution of values ... A true revolution of values will soon cause us to question the fairness and justice of many of our past and present policies.
—**Martin Luther King Jr.**

We're attracted to tales of heroes and villains, of good guys and bad guys. Why wouldn't we be? They're exciting stories, easy to understand. In the story of the "decade we almost stopped climate change," we can identify the People Who Did the Wrong Thing and the Ones Who Tried to Fight Back. It's fun to see ourselves on the Right Side. We love the stories of the Chosen One, and each of us might secretly dream at some point that the Chosen One could be me! We want to be better than we are. We want to be better than one another.

> What can we do to improve ourselves? Of course, we can resist acting on our nastier hierarchical tendencies. Most of us do that most of the time already . . . Tolerance, like any aspect of peace, is forever a work in progress, never completed, and, if we're as intelligent as we like to think we are, never abandoned.
> —Octavia Butler

But . . . what if we're all heroes and villains at the same time? Think about it. Do you do all good things or all bad? (Come on, seriously.) Does there really have to be "The One"? Instead of a Chosen One, what if we choose one another? What if we stop thinking about those sides, about ourselves, and start thinking about one another? What if, instead of thinking of winning and losing, we focus on the ways that we are responsible to and for one another, and what justice means for us all? What if it's about what we've done, what we do, and what we'll do next . . . but also about who we are and who we want to be? If we start by speaking truth to power, with love, maybe we can understand that when we speak about things like fuel-efficiency standards or gasoline taxes or methane flaring or carbon offset credits, about Green and Blue New Deals and bright pink coral, we're not speaking about plans, or programs, or initiatives. We're speaking about a global metamorphosis, about deciding as a people that we can be more and do better and love well. We're speaking about nothing less than all we love and all we are.

Notes

Part I: Where We Are

1 "Today, almost nine": Nathaniel Rich, *Losing Earth: A Recent History* (New York: Farrar Straus Giroux, 2018), pp. 3–4.

CHAPTER 1: LETTING OLD HABITS DIE, HARD

4 "achieve zero-waste schools": "Our Story," *Cafeteria Culture*, n.d., http://www
.cafeteriaculture.org/our-story.html

8 "They keep talking": "Uganda's First Fridays for Future Climate Striker, Vanessa Nakate, Joins COP25 Protests in Madrid," *Democracy Now!*, December 12, 2019, https://www
.democracynow.org/2019/12/12/cop25_vanessa_nakate_uganda

9 "It is when": Kera Sherwood-O'Regan, "Rangatahi Take the UN . . . Again," *The Spinoff*, July 11, 2018, https://thespinoff.co.nz/atea/11–07–2018/rangatahi-take-the-un-again/

9 "My Government has failed": Chloe Farand, "Nine-year-old Girl Files Lawsuit against Indian Government Over Failure to Take Ambitious Climate Action," *The Independent*, April 1, 2017, https://www.independent.co.uk/climate-change/news/nine-ridhima
-pandey-court-case-indian-government-climate-change-uttarakhand-a7661971.html

9 "For young indigenous women": Rayanne Cristine Maximo Franca, "From Where I Stand: 'It Is Time that the World Hears Our Voice,'" *UN Women*, August 4, 2017, https://
www.unwomen.org/en/news/stories/2017/8/from-where-i-stand-rayanne-cristine
-maximo-franca

10 "I started my activism": Brianna Fruean and others, "Young Climate Activists Around the World: Why I'm Striking Today," *The Guardian*, March 15, 2019, https://www
.theguardian.com/commentisfree/2019/mar/15/young-climate-activists-striking
-today-campaigners?CMP=share_btn_link

10 "One lesson much too well": Winnie Asiti, "8,000 Trees in Mt. Kenya, One Year Later,"

AYICC Kenya, July 2010, http://ayicckenya.blogspot.com/2010/07/8000-trees-in
-mt-kenya-one-year-later.html

12 "Your carbon footprint": "What Can We Do to Help?" *NASA Climate Kids*, n.d., https://
climatekids.nasa.gov/review/how-to-help/

13 "These strikes are happening": Eric Holthaus, "Ilhan Omar's 16-year-old Daughter Is Co-
Leading the Youth Climate Strike," *Grist*, March 13, 2019, https://grist.org/article
/ilhan-omars-16-year-old-daughter-is-co-leading-the-youth-climate-strike/

Chapter 2: What Is Climate Change, and Where Did It Come From?

15 "The gargantuan threat": "Young Indigenous Nepalese Woman to Speak at the UN Climate
Summit," Asia Pacific Forum on Women, Law and Development, September 21, 2014,
www.apwld.org/young-indigenous-nepalese-woman-un-climate-summit/

16 "The atmosphere acts": Waldemar Kaempffert, "Warmer Climate on the Earth May Be
Due to More Carbon Dioxide in the Air," *New York Times*, October 28, 1956, p. 191,
https://nyti.ms/3nNq3v3

18 "The atmosphere admits": John Tyndall, "On the Transmission of Heat of Different Qual-
ities Through Gases of Different Kinds," *Pharmaceutical Journal* (2nd ser., 1), December 1859, p. 338.

19 "The increase of": Roger Revelle and Hans E. Suess, "Carbon Dioxide Exchange Between
Atmosphere and Ocean and the Question of an Increase of Atmospheric CO_2, During
the Past Decades," *Tellus* (vol. 9, no. 1), 1957, p. 18.

19 "Since the start": "Science: One Big Greenhouse," *Time*, May 28, 1956, http://content
.time.com/time/subscriber/article/0,33009,937403,00.html

23 "The invasion of the Americas": Mark Trahant, "How Colonization of the Americas Killed
90 Percent of Their Indigenous People—and Changed the Climate," *YES! Magazine*, February 13, 2019, https://www.yesmagazine.org/opinion/2019/02/13/how-colonization
-of-the-americas-killed-90-percent-of-their-indigenous-people-and-changed-the-climate

23 "The regrowth soaked": Oliver Milman, "European colonization of Americas killed
so many it cooled Earth's climate," The Guardian, January 31, 2019, https://www
.theguardian.com/environment/2019/jan/31/european-colonization-of-americas
-helped-cause-climate-change

24 "This new study": Jonathan Amos, "America Colonisation 'Cooled Earth's Climate,'"
BBC News, January 31, 2019, https://www.bbc.com/news/science-environment
-47063973

25 "I love to think": George Washington Carver, *George Washington Carver: In His Own Words* (Columbia: University of Missouri Press, 1987), p. 143, https://archive.org/details /georgewashington00carv/page/142/mode/2up

25 "Without substantial collective": Jeff Tollefson, "COVID curbed carbon emissions in 2020—but not by much," Nature, January 15, 2021, https://www.iea.org/articles /global-energy-review-co2-emissions-in-2020

27 "Greenhouse gases are accumulating": *Climate Change Science: An Analysis of Some Key Questions*, report of the Committee on the Science of Climate Change, Division on Earth and Life Studies. National Research Council (Washington, D.C.: National Academy Press, 2001), p. 1.

27 "that the scientific understanding": "Is Global Warming a Myth?" *Scientific American*, April 8, 2009, https://www.scientificamerican.com/article/is-global-warming-a-myth/

28 "extremely likely to have": Intergovernmental Panel on Climate Change, *Climate Change 2014: Synthesis Report; Contribution of Working Groups I, II and III to the Fifth Assessment Report of the Intergovernmental Panel on Climate Change*, eds. R. K. Pachauri and L. A. Meyer (Geneva: IPCC, 2014), p. 6.

32 "We cannot continue": Annie Leonard and Tom Newmark, "The Story of Soil Is the Story of All of Us," *YES! Magazine*, April 22, 2019, https://www.yesmagazine.org/opinion /2019/04/22/soil-story-community-earth-annie-leonard-tom-newmark

33 "Weather is like": Katharine Hayhoe, "Five Myths About Climate Change," *The Washington Post*, November 30, 2018, https://www.washingtonpost.com/outlook/five -myths/five-myths-about-climate-change/2018/11/30/9fba233a-f428–11e8 -bc79–68604ed88993_story.html

CHAPTER 3: NO JUSTICE, NO PEACE

35 "[Environmental justice is]": Oliver Milman, "Robert Bullard: 'Environmental Justice Isn't Just Slang, It's Real," *The Guardian*, December 20, 2018, https://www.theguardian .com/commentisfree/2018/dec/20/robert-bullard-interview-environmental-justice -civil-rights-movement

35 "The rich got rich": Tom Bawden, "Paris Climate Change Talks: Lord Stern Calls on Rich Countries to Help Poor Nations Cope with Global Warming," *The Independent*, November 24, 2015, https://www.independent.co.uk/climate-change/news/paris -climate-change-talks-lord-stern-calls-on-rich-countries-to-help-poor-nations-cope -with-global-warming-a6747466.html

37 "If it means": Bob Drogin, "Toxic Dirt Dumping Facing Opposition," *The Washington Post*,

January 6, 1979, https://www.washingtonpost.com/archive/politics/1979/01/06
/toxic-dirt-dumping-facing-opposition/2f2723fe-d0c0-464d-b252-eb4040856f32/

38 "environmental racism": Drogin, "Toxic Dirt Dumping Facing Opposition."

38 "more powerful than": Robert D. Bullard, Paul Mohai, Robin Saha, and Beverly Wright,
Toxic Wastes and Race at Twenty: 1987–2007 (Cleveland: United Church of Christ, 2007),
p. x, https://www.nrdc.org/resources/toxic-wastes-and-race-twenty-1987-2007

38 "Americans continually": Rajit Iftikhar, "Amazon Refuses to Act on Climate Change.
So We Employees Are Speaking Out," *YES! Magazine*, May 30, 2019, https://www
.yesmagazine.org/opinion/2019/05/30/amazon-climate-change-action-jeff-bezos

39 "climate refugees": Bruna Kadletz, "Climate Refugees: People and the Environment
Treated as Disposable," *The New Humanitarian*, December 16, 2016, https://deeply
.thenewhumanitarian.org/refugees/community/2016/12/16/climate-refugees
-people-and-the-environment-treated-as-disposable

39 "This is a conversation": Vanessa Nakate, "Uganda's Vanessa Nakate Says COP26 Sidelines
Nations Most Affected by Climate Change," interview by Ari Shapiro, *All Things Considered*, NPR, November 10, 2021, https://www.npr.org/2021/11/10/1053943770
/ugandas-vanessa-nakate-says-cop26-climate-summit-sidelines-global-south

40 "Do whatever you can": Amy Goodman, "We Are Not Responsible": Youth Climate Activists
Rally in Glasgow to Demand World Leaders Act Now," *Democracy Now!*, November 8,
2021, https://www.democracynow.org/2021/11/8/youth_speak_out_at_cop26

42 "A Race without the knowledge": Charles C. Seifert, *The Negro's or Ethiopian's Contribution
to Art* (New York: The Ethiopian Historical Publishing Co., 1938), p. 5.

Part II: Voices Crying Out in the Wilderness

43 "That we came": Rich, *Losing Earth*, p. 9.

CHAPTER 4: A HISTORY OF PROTEST

46 "A chopped head": Blaine O'Neill, "Bishnoi Villagers Sacrifice Lives to Save Trees, 1730,"
Global Nonviolent Action Database, Swarthmore College, December 12, 2010, https://
nvdatabase.swarthmore.edu/content/bishnoi-villagers-sacrifice-lives-save-trees-1730

47 "That is a tree of God": Wangari Maathai, "Marching with Trees," interview by Krista Tip-
pett, *On Being*, NPR, April 6, 2006, https://onbeing.org/programs/wangari-maathai
-marching-with-trees/

48 "are able to go": Maathai, "Marching with Trees."

48 "having learned how": Maathai, "Marching with Trees."

49 "Peace on earth": Nobel Committee, "The Nobel Peace Prize for 2004," *The Nobel Prize*, October 8, 2004, https://www.nobelprize.org/prizes/peace/2004/press-release/

51 "Environment is all": Gaylord Nelson, Earth Day speech, Denver, April 22, 1970, http://www.nelsonearthday.net/docs/nelson_26–18_ED_denver_speech_notes.pdf

53 "I got hooked": "Meet Robert Bullard, the Father of Environmental Justice," interview with Gregory Dicum. *Grist*, March 15, 2006, https://grist.org/article/dicum/

54 "to fight the destruction": For the seventeen principles adopted at the First People of Color Environmental Leadership Summit, Washington, D.C., October 27, 1991, see http://web.archive.org/web/20120219203204/http:/www.weact.org/AboutUs/PrinciplesofEnvironmentalJustice/tabid/226/Default.aspx

54 "was a turning": Jeff Chang [@zentronix], 2021. SolidarityIs . . . fighting to protect the water, air, and land for all. #SolidarityIs . . . the fight for #environmental justice. Instagram, May 29. https://www.instagram.com/tv/CPdswJPhMOw/?utm_medium=copy_link

55 "People get uncomfortable": Dicum, "Meet Robert Bullard."

55 "We sometimes forget": Sophie Hirsh, "The Black Climate Scientists and Scholars Changing the World," *Green Matters*, August 18, 2020, https://www.greenmatters.com/p/black-climate-scientists

56 "A report by": Ken T. Budden and Werner H. Zieger, *Environmental Assessment of Coal Liquefaction: Annual Report* (Research Triangle Park, N.C.: Industrial Environmental Research Laboratory, 1978), p. 66.

CHAPTER 5: SOUNDING THE ALARM

62 "A wait-and-see": Rich, *Losing Earth*, p. 36.

63 "We are facing": Rich, *Losing Earth*, p. 41.

65 "Any government action": *Losing Earth*, p. 41.

CHAPTER 6: ON THE BRINK

67 "the enormous quantity": Rich, *Losing Earth*, p. 48.

68 "When the temperature does": Rich, *Losing Earth*, p. 48.

68 "until at least": Rich, *Losing Earth*, p. 49.

70 "Would anyone like": Rich, *Losing Earth*, p. 55.

74 "permanently and disastrously": Rich, *Losing Earth*, p. 66.

75 "If Gordon MacDonald": Rich, *Losing Earth*, p. 68.

Chapter 7: From Reaction to No Action

78 "He is probably": "The Nobel Peace Prize for 2007," *The Nobel Prize*, October 12, 2007, https://www.nobelprize.org/prizes/peace/2007/press-release/

78 "Believe in the power": Al Gore, "The Most Important Truth. The Power of Your Own Voice," interview with Maranda Pleasant, *Origin Magazine* (issue 14), September 1, 2013, p. 65.

79 "broad consensus": Rich, *Losing Earth*, p. 74.

80 "The trend is all": Rich, *Losing Earth*, p. 79.

81 "extreme negative speculations": Rich, *Losing Earth*, p. 91.

81 "A panel of top": Rich, *Losing Earth*, p. 93.

81 "There are no actions": Rich, *Losing Earth*, p. 94.

83 "It is time to act": Rich, *Losing Earth*, p. 96.

84 "This issue is": Rich, *Losing Earth*, p. 97.

Chapter 8: A Hole in the Sky Changes Everything

86 "The layer is critical": Pawan Bhartia, "Discovering the Ozone Hole: Q & A with Pawan Bhartia," interview by Kathryn Hansen, *NASA Global Climate Change*, September 17, 2012, https://climate.nasa.gov/news/781/discovering-the-ozone-hole-qa-with-pawan-bhartia/

89 "As a reversal": Rich, *Losing Earth*, p. 106.

89 "the costs and benefits": Rich, *Losing Earth*, p. 106.

Chapter 9: Running Out of Time

92 "Before I begin": Rich, *Losing Earth*, p. 118.

93 "At that point": Rich, *Losing Earth*, p. 126.

94 "Will this madness": Rich, *Losing Earth*, p. 127.

94 "The global warming": Rich, *Losing Earth*, p. 130.

96 "with 99 percent confidence": Rich, *Losing Earth*, p. 132.

96 "It is time to stop waffling": Rich, *Losing Earth*, p. 132.

97 "I am an environmentalist": Rich, *Losing Earth*, p. 138.

97 "The drought highlighted": The Bentsen–Quayle Vice Presidential Debate, October 5, 1988, Debate Transcripts, The Commission on Presidential Debates, https://www.debates.org/voter-education/debate-transcripts/october-5–1988-debate-transcripts/

98 "greatly exceed the capacity": Rich, *Losing Earth*, p. 140.

CHAPTER 10: PUSHBACK

100 "an active participant": Rich, *Losing Earth*, p. 148.

100 "Many people are already": Rich, *Losing Earth*, p. 148.

101 "Leave the science to the scientists": Rich, *Losing Earth*, p. 150.

101 "old engineer": Rich, *Losing Earth*, p. 150.

102 "technical garbage": Rich, *Losing Earth*, p. 152.

104 "I should be allowed": Rich, *Losing Earth*, p. 157.

105 "Why do you directly": Rich, *Losing Earth*, p. 157.

106 "I think they're scared": Rich, *Losing Earth*, p. 159.

107 "five levels down": Rich, *Losing Earth*, p. 159.

107 "an outrageous assault": Rich, *Losing Earth*, p. 159.

107 "a cold war": Rich, *Losing Earth*, p. 159.

107 "the signal that": Rich, *Losing Earth*, p. 159.

107 "to develop full": Rich, *Losing Earth*, p. 162.

108 "Once again, the president": Rich, *Losing Earth*, p. 162.

108 "I don't want anyone": Rich, *Losing Earth*, p. 163.

CHAPTER 11: MISSED OPPORTUNITY

112 "forceful and specific": Rich, *Losing Earth*, p. 167.

113 "play a leadership": Rich, *Losing Earth*, p. 167.

113 "Who picks up": Rich, *Losing Earth*, p. 168.

116 "whitewash effect": Rich, *Losing Earth*, p. 171.

116 "The president made": Rich, *Losing Earth*, p. 171.

Part III: Handling the Truth

117 "For a long time": Avichai Scher, "'Climate Grief': The Growing Emotional Toll of Climate Change," *NBC News*, December 24, 2018, https://www.nbcnews.com/health/mental-health/climate-grief-growing-emotional-toll-climate-change-n946751

117 "We may encounter": Marianne Schnall, "An Interview with Maya Angelou," *Psychology*

Today, February 17, 2009, psychologytoday.com/us/blog/the-guest-room/200902
/interview-maya-angelou.

Chapter 12: That Was Then, This Is Now (And Then)

119 "Yes . . . And no.": Rich, *Losing Earth*, p. 177.

120 "The leaders in": Rich, *Losing Earth*, p. 179.

120 "Who knows?": Rich, *Losing Earth*, p. 179.

121 "premature" policies: Rich, *Losing Earth*, p. 182.

123 "You know very well": Rich, *Losing Earth*, p. 184.

123 "considerable doubt": Rich, *Losing Earth*, p. 184.

123 "outspoken greenhouse dissidents": Rich, *Losing Earth*, p. 184.

126 "We didn't want": Rich, *Losing Earth*, p. 186

128 "Coal and oil": Rich, *Losing Earth*, p. 189.

128 "man's factory chimneys": Rich, *Losing Earth*, p. 190.

129 "Man may be": Rich, *Losing Earth*, p. 190.

130 "I believe man": Rich, *Losing Earth*, p. 195.

130–131 "We don't understand": Rich, *Losing Earth*, p. 196.

131 "I would not": Tom DiChristopher, "EPA Chief Scott Pruitt Says Carbon Dioxide Is Not
a Primary Contributor to Global Warming," *CNBC*, March 9, 2017, https://www
.cnbc.com/2017/03/09/epa-chief-scott-pruitt.html

131 "I think there's": Donald Trump, interview by *The Washington Post* editorial board,
The Washington Post, March 21, 2016, https://www.washingtonpost.com/blogs
/post-partisan/wp/2016/03/21/a-transcript-of-donald-trumps-meeting-with-the
-washington-post-editorial-board/

132 "points to the loss": Rich, *Losing Earth*, p. 199.

132 "Anybody ever turn": "Remarks by President Biden on the Bipartisan Infrastructure Framework," The White House, June 29, 2021, https://www.whitehouse.gov
/briefing-room/speeches-remarks/2021/06/29/remarks-by-president-biden-on-the
-bipartisan-infrastructure-framework/

133 "first major investment": Coral Davenport and Christopher Flavelle, "Infrastructure Bill
Makes First Major U.S. Investment in Climate Resilience," *New York Times*, November 6,
2021, https://www.nytimes.com/2021/11/06/climate/infrastructure-bill-climate.html

133 "It's significant that": Davenport and Flavelle, "Infrastructure Bill Makes First Major U.S. Investment in Climate Resilience."

134 "We're still at": James Hansen, interview by Bob McDonald, CBC Radio, May 12, 2018, https://www.cbc.ca/radio/quirks/may-12-2018-james-hansen-s-i-told-you -so-elephant-earthquakes-and-third-hand-smoke-1.4656916

Chapter 13: Time for Change

135 "a small group": Donald Keys, *Earth at Omega: Passage to Planetization* (Boston: Branden Press, 1983), page 79.

136 "We are at": Xiuhtezcatl Roske-Martinez, "Xiuhtezcatl, Indigenous Climate Activist at the High-level event on Climate Change," United Nations, YouTube video, 9:31, June 29, 2015, https://www.youtube.com/watch?v=27gtZ1oV4kw

136 "Our ancestors are": Jaskiran Dhillon, "Indigenous Youth Are Building a Climate Justice Movement by Targeting Colonialism," *Truthout*, June 20, 2016, https://truthout.org /articles/indigenous-youth-are-building-a-climate-justice-movement-by-targeting -colonialism/

137 "The Indigenous Peoples": "Mission, Principles and Code of Ethics," Indigenous Environmental Network, n.d., https://www.ienearth.org/about/

138 "This is an environmental": "Quick Facts & Main Concerns About the Dakota Access Pipeline," Grassroots Global Justice Alliance, n.d., https://ggjalliance.org/resources /quick-facts-main-concerns-about-the-dakota-access-pipeline/

138 "Indigenous peoples are living": Dhillon, "Indigenous Youth Are Building."

139 "Women were the people": Dhillon, "Indigenous Youth Are Building."

140 "We must not sit": "Call to Action this World Environment Day," *Panafrican Climate Justice Alliance*, June 4, 2020, http://oldsite.pacja.org/category/item/829-call-to-action-this -world-environment-day%20128

140 "A lack of political": Carol Rasmussen, "Just 5 Questions: Community Initiatives Against Climate Change," *NASA Global Climate Change*, January 9, 2014, https://climate.nasa .gov/news/1026/just-5-questions-community-initiatives-against-climate-change/

141 "Beginning in 1998": Rasmussen, "Just 5 Questions."

142 "Our house is": 350.org, "Greta Thunberg at the Global Climate Strike in New York City," YouTube Video, 9:47, September 23, 2019, https://www.youtube.com/watch ?v=tALlM6uUWrc

144 "recognizing the duty": "House Resolution 109—Recognizing the duty of the Federal

Government to create a Green New Deal," 116th Congress (2019–2020), February 7, 2019. Congress.Gov, https://www.congress.gov/bill/116th-congress/house-resolution/109/text

144 "Because the United States": "House Resolution 109."

145–146 "The Green New Deal": "Green New Deal," Sunrise Movement, n.d., https://web.archive.org/web/20191006011825/https://www.sunrisemovement.org/green-new-deal

146 "Growing up I'd wonder": Donna M. Owens, "She, The People: Meet Rhiana Gunn-Wright, An Architect Behind The Green New Deal," *Essence*, April 17, 2019, https://www.essence.com/feature/she-the-people-rhiana-gunn-wright-green-new-deal/

146 "meeting 100 percent": "House Resolution 109."

146 "This is going to be": Alexandria Ocasio-Cortez, remarks at "Solving Our Climate Crisis," National Town Hall, Washington, D.C., December 3, 2018, https://www.c-span.org/video/?455281-1/senator-bernie-sanders-holds-town-hall-meeting-climate-change

147 "Overfishing, pollution": Ayana Elizabeth Johnson, "The Key to Halting Climate Change: Admit We Can't Save Everything," *The Guardian*, February 17, 2016, https://www.theguardian.com/commentisfree/2016/feb/17/climate-change-admit-we-cant-save-everything

150 "The avalanche of plastics": "Sweeping New Report on Global Environmental Impact of Plastics Reveals Severe Damage to Climate," Center for International Environmental Law, May 15, 2019, https://www.ciel.org/news/plasticandclimate/

151 "For two and a half ": Fran Korten, "Where Kids Fought Plastic Pollution—and Won," *YES! Magazine*, March 20, 2019, https://www.yesmagazine.org/environment/2019/03/20/kids-get-plastic-ban-ocean-pollution-climate

151 "Know that your voice": "Kristal Ambrose's Acceptance Speech, 2020 Goldman Environmental Prize," Goldman Environmental Prize, YouTube video, 1:10, November 30, 2020, https://www.youtube.com/watch?v=qLOi3hdPHds

152 "insurance policy": "Svalbard Global Seed Vault," Crop Trust, n.d., croptrust.org/our-work/svalbard-global-seed-vault

154 "Scientists would just take": Erin Sagen, "Indigenous Seed Savers Gather in the Andes, Agree to Fight Climate Change with Biodiversity," *YES! Magazine*, July 31, 2014, yesmagazine.org/environment/2014/07/31/indigenous-seed-savers

155 "Climate change will": "Sacred Valley of the Incas to Send Potatoes to Seed Vault," press

release, Crop Trust, February 5, 2011, https://www.croptrust.org/press-release /sacred-valley-incas-send-potatoes-seed-vault/

155 "it's time traditional": Sagen, "Indigenous Seed Savers."

Chapter 14: The Climate Justice Movement

156 "Climate action can't": Eric Holthaus, "The story of us: reimagining the climate narra- tive," *The Correspondent*, July 2, 2020, https://thecorrespondent.com/565/the-story -of-us-reimagining-the-climate-narrative

157 "Cut carbon pollution": "The President's Climate Action Plan," Executive Office of the President, June 2013, https://obamawhitehouse.archives.gov/sites/default/files /image/president27sclimateactionplan.pdf

158 "Reducing emissions": "Climate Mobilization Act," New York City Council, n.d., https://council.nyc.gov/data/green/#green-bills

158 "cities will be home": Jason Plautz, "UN Report: Cities Are 'Key Implementers' of Cli- mate Policies," Smart Cities Dive, December 11, 2018, https://www.smartcitiesdive .com/news/un-report-cities-are-key-implementers-of-climate-policies/544041/

158 "I think the message": Mary Annaïse Heglar, "It's More Fixable Than We Thought with Kate Marvel," *Hot Take*, August 22, 2021, https://hot-take.ghost.io/more-fixable-than -we-thought/

159 "to be the most": "Questions to the President of COP26," *Alok Sharma MP*, June 9, 2021, https://www.aloksharma.co.uk/parliament/questions-president-cop26-1

159 "COP26 is a performance": Libby Brooks, "Hundreds of Global Civil Society Representatives Walk Out of COP26 in Protest," *The Guardian*, November 21, 2021, https://www.theguardian.com/environment/2021/nov/12/global-civil-society -representatives-walkout-cop26-protest?CMP=twt_a-environment_b-gdneco

159 "At least 503 fossil fuel lobbyists": "Hundreds of Fossil Fuel Lobbyists Flooding COP26 Climate Talks," *Global Witness*, November 8, 2021, https://www.globalwitness.org/en /press-releases/hundreds-fossil-fuel-lobbyists-flooding-cop26-climate-talks/

160 "The presence of": "Hundreds of Fossil Fuel Lobbyists Flooding COP26 Climate Talks."

160 "two largest economies": Fiona Harvey, "China and the US Announce Plan to Work Together on Cutting Emissions," *The Guardian*, November 10, 2021, https://www .theguardian.com/environment/2021/nov/10/china-and-the-us-announce-plan-to -work-together-on-cutting-emissions?CMP=Share_iOSApp_Other

160 "attempts to ensure": https://www.reuters.com/business/cop/cop26-publishes-new -draft-declaration-kicking-off-more-horse-trading-2021-11-12/

161 "The concept of global warming": Louis Jacobson, "Yes, Donald Trump did call climate change a Chinese hoax," *PolitiFact*, June 3, 2016, https://www.politifact.com/factchecks /2016/jun/03/hillary-clinton/yes-donald-trump-did-call-climate-change-chinese-h/.

161 "I think the climate": Jacobson, "Yes, Donald Trump did."

161 "Something's changing": "President Trump on Christine Blasey Ford, his relationships with Vladimir Putin and Kim Jong Un and more," interview with Lesley Stahl, *60 Minutes*, October 15, 2018, https://www.cbsnews.com/news/donald-trump-full -interview-60-minutes-transcript-lesley-stahl-2018–10–14/

162–163 "Chinese virus": Kimmy Yam, "Trump Doubles Down That He's Not Fueling Racism, But Experts Say He Is," *NBC News*, March 18, 2020, https://www.nbcnews.com/news /asian-america/trump-doubles-down-he-s-not-fueling-racism-experts-say-n1163341

163 "Because we're far more": Mary Annaïse Heglar, "We Don't Have To Halt Climate Action To Fight Racism," *Huffington Post*, June 16, 2020, https://www.huffpost.com/entry /climate-crisis-racism-environmenal-justice_n_5ee072b9c5b6b9cbc7699c3d

164 "Oil, gas, and petrochemical": Rhiana Gunn-Wright, "Think This Pandemic Is Bad? We Have Another Crisis Coming," *The New York Times*, April 15, 2020, https://www.nytimes .com/2020/04/15/opinion/sunday/climate-change-covid-economy.html

164 "consideration of a green": Justin Worland, "As the Rest of the World Plans a Green Recovery, America Is Once Again Falling Behind," *Time*, May 15, 2020, https://time .com/5835402/green-stimulus-climate-change-coronavirus/

164 "A lax definition": Gunn-Wright, "Think This Pandemic Is Bad?"

165 "Instead of leading": Worland, "As the Rest of the World Plans."

166 "Racial–ethnic disparities": Christopher W. Tessum, Joshua S. Apte, Andrew L. Good-kind, Nicholas Z. Muller, Kimberley A. Mullins, David A. Paolella, Stephen Polasky, et al., "Inequity in Consumption of Goods and Services Adds to Racial–Ethnic Dispari-ties in Air Pollution Exposure," *Proceedings of the National Academy of Sciences of the United States of America*, March 26, 2019 (vol. 116, no. 13), pp. 6001–6006; https://www .pnas.org/content/116/13/6001

167 "We can no longer": "Remarks by President Biden at the 2021 Virtual Munich Security Conference," The White House, February 19, 2021, https://www.whitehouse.gov /briefing-room/speeches-remarks/2021/02/19/remarks-by-president-biden-at-the -2021-virtual-munich-security-conference/

Chapter 15: Tomorrow People

169 "Tomorrow belongs only": "Malcolm X's Speech at the Founding Rally of the Organization of Afro-American Unity," *Black Past*, October 15, 2007, https://www.blackpast.org/african-american-history/speeches-african-american-history/1964-malcolm-x-s-speech-founding-rally-organization-afro-american-unity/

169 "The truth is": Wanjiku Gatheru, "Want to Be an Environmentalist? Start with Anti-racism," *Glamour*, July 30, 2020, https://www.glamour.com/story/want-to-be-an-environmentalist-start-with-anti-racism

170 "It's not just time": Heglar, "We Don't Have To Halt Climate Action."

170 "from the time": Tansy Kaschak, "Landscapes of Color," *Open Spaces* (1) n.d., p. 126.

170 "Public lands belong": Jose G. Gonzalez, "¡Estamos Aquí! Opening America's Public Lands and Green Spaces," *Huffington Post*, November 3, 2016, https://www.huffpost.com/entry/estamos-aqu%C3%AD-opening-americas-public-lands-and-green_b_581b6f33e4b0d68ecbca8ed4?timestamp=1478195213001

171 "You can be a kid": Justine Calma and Paola Rosa-Aquino, "4 Black Women Leaders on Climate, Justice, and the Green 'Promised Land,'" *Grist*, February 27, 2019, https://grist.org/article/4-black-women-leaders-on-climate-justice-and-the-green-promised-land/

171 "I am inspired": Calma and Rosa-Aquino, "4 Black Women Leaders on Climate."

171 "Getting young people out": Eric Holthaus, "Ilhan Omar's 16-Year-Old Daughter Is Co-Leading the Youth Climate Strike," *Grist*, March 19, 2019, https://grist.org/article/ilhan-omars-16-year-old-daughter-is-co-leading-the-youth-climate-strike/

172 "It's important to": Holthaus, "Ilhan Omar's 16-Year-Old Daughter."

173 "inclusive groups": Niellah Arboine, "How to Fight for Climate Action, According to Isra Hirsi," *Dazed*, May 13, 2019, https://www.dazeddigital.com/politics/article/44390/1/how-to-fight-for-climate-action-isra-hirsi-climate-strike-ilhan-omar

173 "For us, it's about": Mike Seely, "As Students Clamor for More on Climate Change, Portland Heeds the Call," *The New York Times*, June 9, 2019, https://www.nytimes.com/2019/06/09/us/portland-climate-change.html

174 "I remember thinking": Greta Thunberg, translated by Oyama Akinori, "The Disarming Case to Act Right Now on Climate Change," TEDxStockholm, November 2018, https://www.ted.com/talks/greta_thunberg_the_disarming_case_to_act_right_now_on_climate_change/transcript?language=en

176 "How dare you!": Guardian News, "Greta Thunberg to world leaders: 'How dare you?

You have stolen my dreams and my childhood,' YouTube Video, 4:34, September 23, 2019, https://www.youtube.com/watch?app=desktop&v=TMrtLsQbaok

176 "We know what": https://listen.sdpb.org/post/greta-thunberg-and-tokata-iron-eyes-talk-keystone-xl-and-climate-change

176 "This crisis does not": https://ndncollective.org/thunberg-and-takota-iron-eyes-join-forces-and-elevate-climate-justice-conversation/

177 "at the edge": Natasha Rausch, "Swedish activist Greta Thunberg brings climate message to Standing Rock Sioux Nation," *The Jamestown Sun*, October 8, 2019, https://www.jamestownsun.com/news/government-and-politics/4711602-Swedish-activist-Greta-Thunberg-brings-climate-message-to-Standing-Rock-Sioux-Nation

177 "noticed that my classmates": Xiye Bastida, "If You Adults Won't Save the World, We Will," TED2020, July 2020, https://www.ted.com/talks/xiye_bastida_if_you_adults_won_t_save_the_world_we_will/transcript?language=en

178 "As much as": Kenya Evelyn, "'Like I Wasn't There': Climate Activist Vanessa Nakate on Being Erased from a Movement," *The Guardian*, January 29, 2020, https://www.theguardian.com/world/2020/jan/29/vanessa-nakate-interview-climate-activism-cropped-photo-davos

179 "I have been talking": Ciarra Jones, "Climate Activist Vanessa Nakate Wants a More Inclusive Movement in 2021," *Elite Daily*, December 9, 2020, https://www.elitedaily.com/p/climate-activist-vanessa-nakate-wants-a-more-inclusive-movement-in-2021-48680096

179 "I see my role": Vanessa Nakate, "Vanessa Nakate Wants Climate Justice for Africa," *Time*, October 28, 2021, https://time.com/6109452/vanessa-nakate-climate-justice/

180 "All of us know": Steven Lee Myers, "Ignored and Ridiculed, She Wages a Lonesome Climate Crusade," *New York Times*, December 4, 2020, https://www.nytimes.com/2020/12/04/world/asia/ou-hongyi-china-climate.html

180 "China doesn't need": Sally Ho, "18 Things to Know about Howey Ou, China's Only Teenage Climate Striker," *National Catholic Reporter*, August 25, 2020, https://www.ncronline.org/news/earthbeat/18-things-know-about-howey-ou-chinas-only-teenage-climate-striker

180 "What does climate change": Ted Richane, "Vic Barrett: Youth Activist," UCLA: Institute of the Environment and Sustainability, n.d., https://www.ioes.ucla.edu/person/vic-barrett/

180 "I think imagining": "Meet a Young Activist Who's Suing the Government Over Climate Change," Mashable, December 12, 2020, https://mashable.com/article/vic-barrett-young-activism-intersectionality

181 "I advocate for": Autumn Peltier, "The Powerful Impact of Activism," interview by Izida Zorde, *ETFO Voice*, Spring 2018, https://etfovoice.ca/feature/interview-autumn-peltier

181 "Clean water is needed": Colleen Romaniuk, "Clean Water for First Nations Critical During the COVID-19 Pandemic: Activists," Thunder Bay News Watch, January 10, 2021, https://www.tbnewswatch.com/local-news/clean-water-for-first-nations -critical-during-the-covid-19-pandemic-activists-3240975

181 "I feel that": Chere Di Boscio, "Meet Maya S. Penn, the Keen, Green Wünder Teen," *Eluxe Magazine*, July 4, 2013, https://eluxemagazine.com/people/keen-green-and -just-13-maya-shea-penn/

182 "where ethically problematic companies": Laura Pitcher, "Woke-Washing: What Is It and How Does It Affect the Fashion Industry?" *Teen Vogue*, November 3, 2021, https://www .teenvogue.com/story/what-is-woke-washing

182 "Let's cancel": Alexandria Villaseñor [@AlexandriaV2005], 2021. Let's cancel the Youth Climate Hope Industrial Complex now, before #COP26. If you're relying on youth to save us, then you're not doing enough yourself. Get in the streets, call your law-makers, phone bank, talk to your friends, families & communities. Twitter. October 25. https://twitter.com/AlexandriaV2005/status/1452677215154364416

182 "has violated the": "Juliana v. United States," Our Children's Trust, n.d., https://www .ourchildrenstrust.org/juliana-v-us

183 "When those in power": Dhillon, "Indigenous Youth."

183 "The plaintiffs' case": Opinion by Judge Andrew D. Hurwitz, dissent by Judge Jo-sephine L. Staton, *Juliana v. United States*, United States Court of Appeals for the Ninth Circuit, filed January 17, 2020, https://static1.squarespace.com/static /571d109b04426270152febe0/t/5e22101b7a850a06acdff1bc/1579290663460/2020 .01.17+JULIANA+OPINION.pdf

184 "They have consistently said": Rachel McDonald, "Settlement Negotiations Fail Be-tween Climate Kids and Government Attorneys," *KLCC*, November 1, 2021, https:// www.klcc.org/crime-law-justice/2021-11-01/settlement-negotiations-fail-between -climate-kids-and-government-attorneys

Chapter 16: Give Light, and People Will Find a Way

186 "How does a sentient": Rich, *Losing Earth*, pp. 192–193.

186 "Some Americans already are": Kevin J. Coyle and Lise Van Susteren, *The Psychological Effects of Global Warming on the United States*, National Forum and Research Report,

February 2012, https://fdocuments.us/document/the-psychological-effects-of-global-warming-on-the-united-states.html

186 "People can't fix": Mary Annaïse Heglar, "Climate Grief Hurts Because It's Supposed To," *The Nation*, November 7, 2021, https://www.thenation.com/article/environment/climate-grief-hope/

187 "A lot of what": Jeremy Deaton, "Uncovering the Mental Health Crisis of Climate Change," Nexus Media News, June 12, 2018, https://nexusmedianews.com/uncovering-the-mental-health-crisis-of-climate-change-88cd35a1293c/

187 "the little wins": Aditi Murti, "Closeness to Climate Change and Pollution Can Cause Distress, Environmental Grief," *The Swaddle*, August 12, 2019, https://theswaddle.com/environmental-grief/

187 "ecological grief": Ashlee Cunsolo and Neville R. Ellis, "Ecological Grief as a Mental Health Response to Climate Change-Related Loss," *Nature Climate Change* (vol. 8), April 2018, pp. 275–281, https://doi.org/10.1038/s41558-018-0092-2

187 "climate grief": Scher, "'Climate Grief': The Growing Emotional Toll of Climate Change."

187 "reach across differences": Cunsolo and Ellis, "Ecological Grief as a Mental Health Response to Climate Change-Related Loss," p. 276.

189 "What helps people": Scher, "'Climate Grief': The Growing Emotional Toll of Climate Change."

189 "whether that terrain": Sarah Jaquette Ray, "Climate Anxiety Is an Overwhelmingly White Phenomenon," *Scientific American*, March 21, 2021, https://www.scientificamerican.com/article/the-unbearable-whiteness-of-climate-anxiety/

189 "Doing something every day": Lazarovic, Sarah, "How to Not Be (Completely) Depressed About Climate Change," *YES! Magazine*, January 7, 2019, https://www.yesmagazine.org/environment/2019/01/07/how-to-not-be-completely-depressed-about-climate-change

190 "The soil stewards": Leah Penman, "By Reconnecting With Soil, We Heal the Planet and Ourselves," *YES! Magazine*, February 14, 2019, https://www.yesmagazine.org/issue/dirt/2019/02/14/by-reconnecting-with-soil-we-heal-the-planet-and-ourselves

190 "could seriously make up for": Jackie Flynn Mogensen, "Stop Building a Spaceship to Mars and Just Plant Some Damn Trees," *Mother Jones*, 4 July 2019, https://www.motherjones.com/environment/2019/07/stop-building-a-space-ship-to-mars-and-just-plant-some-damn-trees/

190 "the restoration of trees": Jean-Francois Bastin, Yelena Finegold, Claude Garcia, Danilo Mollicone, Mercelo Rezende, Devin Routh, Constantin M. Zohner, and Thomas

Crowther, "The Global Tree Restoration Potential," *Science* (vol. 365, no. 6448) July 5, 2019, pp. 76–79, https://science.sciencemag.org/content/365/6448/76

191 "Groups are more effective": Cathy Brown, "A Climate Action for Every Type of Activist," *YES! Magazine*, July 17, 2019, https://www.yesmagazine.org/issue/travel/2019/07/17/climate-change-take-action-activist

192 "Activists Remove": Joe McCarthy and Erica Sánchez, "Activists Remove 40 Tons of Plastic Waste From Pacific Ocean," *Global Citizen*, July 1, 2019 https://www.globalcitizen.org/en/content/40-tons-plastic-removed-ocean/?template=next

193 "Climate change isn't": Nadja Sayej, "Artists on Climate Change: The Exhibition Tackling a Global Crisis," *The Guardian*, May 30, 2018, https://www.theguardian.com/artanddesign/2018/may/30/climate-change-exhbition-storm-king-new-york

193 "to cultivate a sense": Zoë Lescaze, "12 Artists On: Climate Change," *New York Times Style Magazine*, August 22, 2018, nytimes.com/2018/08/22/t-magazine/climate-change-art.html.

194 "how the increasing": Lescaze, "12 Artists On: Climate Change."

195 "When people try": Patrik Sörqvist and Linda Langeborg, "Why People Harm the Environment Although They Try to Treat It Well," *Frontiers in Psychology*, March 4, 2019 https://www.frontiersin.org/articles/10.3389/fpsyg.2019.00348/full

195 "We know of course": Arundhati Roy, "Peace and the New Corporate Liberation Theology," Sydney Peace Prize lecture, November 3, 2004, https://sydneypeacefoundation.org.au/wp-content/uploads/2012/02/2004-SPP_-Arundhati-Roy.pdf

196 "This is a call": Nicci Attfield, "Toward Eco-Social Justice," *YES! Magazine*, May 10, 2021, https://www.yesmagazine.org/issue/solving-plastic/2021/05/10/eco-social-justice

196 "The fight for": Ayisha Siddiqa [@Ayishas12], 2021. The fight for climate justice at its core a fight for love, it comes from a place of deep deep love for humanity. For without humanity love ceases to exist, laughter ceases to exist. Every little twig that has created our collective concept of this nest we call home, disappears. Twitter. February 15. https://twitter.com/Ayishas12/status/1361237456088141826

196 "civil rights movement": bell hooks, "Love as the Practice of Freedom," *Outlaw Culture: Resisting Representations* (New York: Routledge, 2006), p. 244, https://archive.org/details/outlawcultureres00hook_0/page/244/mode/2up

197 "Often, the shipments": Shashank Bengal, "Asian Countries Take a Stand against the Rich World's Plastic Waste," *Los Angeles Times*, June 17, 2019, https://www.latimes.com/world/la-fg-asia-plastic-waste-20190617-story.html

197 "In the long term": Yen Nee Lee, "The World Is Scrambling Now That China Is Refusing to Be a Trash Dumping Ground," *CNBC*, April 16, 2018, https://www.cnbc.com/2018/04/16/climate-change-china-bans-import-of-foreign-waste-to-stop-pollution.html

198 "the environmental movement": Mary Annaïse Heglar, "Climate Change Isn't the First Existential Threat," *Zora*, February 18, 2019, https://zora.medium.com/sorry-yall-but-climate-change-ain-t-the-first-existential-threat-b3c999267aa0

199 "I am convinced": Martin Luther King Jr., "Beyond Vietnam," speech delivered April 4, 1967, A Call to Conscience: The Landmark Speeches of Dr. Martin Luther King, Jr, https://ratical.org/ratville/JFK/MLKapr67.pdf

200 "What can we do": Octavia E. Butler, "Essay on Racism," NPR, n.d., https://legacy.npr.org/programs/specials/racism/010830.octaviabutleressay.html

Recommended Resources

Books

Bill Bigelow and Tim Swinehart, eds., *A People's Curriculum for the Earth: Teaching Climate Change and the Environmental Crisis* (Milwaukee, WI: Rethinking Schools, 2014). A valuable collection of stories, articles, role plays, and activities.

Nathaniel Rich, *Losing Earth: A Recent History* (New York: Farrar Straus Giroux, 2019). The inspiration for this book, particularly Part II; an eye-opening portrayal of the birth of climate denialism and the genesis of the fossil fuel industry's coordinated effort to thwart climate policy through misinformation propaganda and political influence.

Vanessa Nakate, *A Bigger Picture: My Fight to Bring a New African Voice to the Climate Crisis* (Boston: Mariner Books, 2021). A memoir and call to action, amplifying African and other stories often marginalized in the narrative of the climate justice movement.

Online Resources

The All We Can Save Project. Offshoot of the *All We Can Save* anthology, edited by Ayana Elizabeth Johnson and Katharine K. Wilkinson, this is a feminist collective that emphasizes a narrative shift toward woman-centered leadership and solutions for the climate crisis. https://allwecansave.earth/

Asian Pacific Environmental Network. With a focus on California's Asian immigrant and refugee communities, works to promote environmental, social, and economic justice. https://apen4ej.org/

Caribbean Youth Environment Network. Young people promoting sustainability and climate justice in the Caribbean. https://cyen.org/

Climate Awakening. Founded by Dr. Margaret Klein Salamon, uses small-group conversations to help people process difficult emotions about the climate crisis. https://climateawakening.org/

Climate Feedback. Scientists fact-check climate news reports and claims. https://climatefeedback.org/

Democracy Now! climate coverage. Frequent and comprehensive coverage of stories, communities, and Indigenous, Black, and Brown activists often ignored by mainstream media outlets. https://democracynow.org/topics/climate_change

Disability-Inclusive Climate Action Research Programme. McGill University–based group of scholars and activists working to promote disability-aware solutions to the climate crisis. https://disabilityinclusiveclimate.org/

Earth Guardians. Intergenerational grassroots organization using art, music, and storytelling for youth leadership training and civic engagement. https://earthguardians.org/

Extinction Rebellion. An international, nonpartisan movement dedicated to using nonviolent direct action to compel governments to work for climate justice around the world. https://rebellion.global/

Fridays for Future. Started by Greta Thunberg, a global youth strike movement and organization working to call attention to the climate crisis. Its occasional publication *Newsletter for Future* includes coverage of discrimination against and exclusion of people with disabilities in the climate justice movement. https://fridaysforfuture.org/

Good Grief Network. Uses a ten-step approach to helping people manage climate grief and anxiety and engage in positive action. https://goodgriefnetwork.org/

Hot Take. Newsletter from Mary Annaïse Heglar and Amy Westervelt with essays, news articles, and conversations focused on climate justice. https://hot-take.ghost.io/

How to Save a Planet. Podcast focused on "energizing" conversations about the positive work we can do in this crisis, and success stories of communities and groups in the environmental justice movement. https://gimletmedia.com/shows/howtosaveaplanet

Indigenous Environmental Network. An organizing and information hub that works for environmental justice through direct action, education, and capacity building in Indigenous communities and tribal governments in North America and around the world. https://ienearth.org/

Intersectional Environmentalist. Founded by Leah Thomas, a community and resource hub centering BIPOC voices in the climate movement. Works to promote climate activism that's grounded in social justice. https://intersectionalenvironmentalist.com/

A Matter of Degrees. Solutions-oriented podcast about the global climate justice movement. https://degreespod.com/

Our Climate Our Future. From Action for the Climate Emergency, a video series about the climate crisis. https://ourclimateourfuture.org/

Re-Earth Initiative. Information tool kits, webinars, digital campaigns, and other resources. https://reearthin.org/

Slow Factory. Works to support, educate, and train people from historically marginalized communities to claim leadership roles in the climate justice movement. Open Edu is their series of free environmental justice courses. https://slowfactory.earth/

Soul Fire Farm. An Afro-Indigenous community farm focused on environmental justice, sustainable agriculture, community health, and training of "activist-farmers." https://soulfirefarm.org/

Teen Vogue climate coverage. Accessible coverage of leading voices and current events with a focus on youth activism. https://teenvogue.com/tag/climate-change

350.org. Named after 350 parts per million—the safe concentration of carbon dioxide in the atmosphere—a nonprofit organization working to move the world away from a dependence on fossil fuels toward renewable energy sources. https://350.org/

YES! Media. Solutions-focused, in-depth, and accessible reporting on environmental issues with an emphasis on community-building and grassroots action. https://yesmagazine.org/topic/environment

Youth Climate Lab. Youth-focused lab working on developing innovative climate action projects. https://youthclimatelab.org/

Voices of Youth tools for young climate activists. From UNICEF, resource guides and other information for learning about the climate crisis around the world. https://voicesofyouth.org/climate-toolkit

Zero Hour. Youth-led movement focused on inclusive organizing and mobilization. http://thisiszerohour.org/

Acknowledgments

Abundant gratitude to:

My editor Wesley Adams, for his extraordinarily thoughtful editing, unwavering encouragement, and faith in my vision for this project; and Melissa Warten for all of her good-natured support throughout. Many thanks to the entire FSG and Macmillan family, including Ilana Worrell, Sherri Schmidt, Maria Vlasak, Linda Minton, John Nora, Mallory Grigg, Kathleen Breitenfeld, Angela Jun, Josh Rubins, and Arc Indexing, Inc., for their attention to detail and incredible care throughout the process.

Nathaniel Rich, for his generous introduction and the powerful reporting and compelling storytelling that inspired this book, and Tim Foley for the dynamic and vibrant interior art.

My agent, Marietta Zacker, for having my back, always.

My sister Kikelomo, Renée Watson, Kelly Starling Lyons, Dhonielle Clayton, Lamar Giles, and the LSG for cheering me on when things seemed impossible.

Anna Allanbrook and Cora Sangree for their help with my research of the Styrofoam Out of Schools story, and the entire Brooklyn New School community for its creativity, ingenuity, and passion for environmental justice.

The ancestors, activists, artists, thinkers, scientists, farmers, writers, dreamers, and people of all ages who have worked and continue to work, so often unseen and unthanked, for climate justice. Thank you, thank you, thank you.

My daughter Adedayo, one-time member of the Junior Drip Patrol, my source of inspiration, for her unconditional love and support. You are extraordinary.

Index